Death Stars,

Weird Galaxies, and a

Quasar-Spangled Universe

Death Stars,
Weird Galaxies, and a
Quasar-Spangled Universe

The Discoveries of the Very Large Array Telescope

Karen Taschek

University of New Mexico Press   Albuquerque

YEAR   PRINTING
12 11 10 09 08 07 06   1 2 3 4 5 6 7

ISBN 13: 978-0-8263-3211-0 (CL.)
ISBN 10: 0-8263-3211-0 (CL.)

LIBRARY OF CONGRESS CATALOGING-IN-PUBLICATION DATA

Taschek, Karen, 1956–
        Death stars, weird galaxies, and a quasar-spangled universe :
the discoveries of the Very Large Array telescope / Karen Taschek.
        p. cm.
  Includes index.
  ISBN 0-8263-3211-0 (alk. paper)
 1. Radio astronomy.
 2. Very Large Array.
 3. Galaxies.
 4. Solar system.
 5. Discoveries in science.
I. Title.
  QB476.5.T37   2006
  522'.682—dc22

2005022841

Book and cover design and
typography by Kathleen Sparkes.

Image and illustration credits on page 75.

This book is typeset using Utopia 11/16.
Caption type is Futura Light 10/14.
Display type is the Futura family.

This book was printed by
TWP, in Singapore.

To my dad,

who taught me to look for

the "beauty of mathematics";

to my son, David, the sun of my life;

and to David Holtby, editor in chief

at the University of New Mexico Press,

for his vision in bringing

science books to children.

Contents

Acknowledgments

Special thanks to Bob Broilo, electrical engineer
at the Very Large Array site; Patricia Smiley,
Information Services Coordinator at the
National Radio Astronomy Observatory;
Eric Lynn, former head of the Nuclear Physics
Division of the United Kingdom Atomic Energy
Authority; Liz Laemmle, Youth Service Librarian
at Los Alamos Mesa Public Library; Thomas
Allen and Cathy Bailey, science teachers at
Bosque Middle School; and Michael Westphal
and Michele Sequeria, materials scientists at
Intel. Their scientific, technical, and children's
literature expertise made this book possible.

A tour group at the VLA site on the Plains of San Agustin, near the town of Socorro, New Mexico.

ONE

A Tour of the Very Large Array

What do you see when you look up at the clear, dark night sky? An explosion of stars sprinkled across black, bottomless outer space—perhaps 5,000 stars are visible. But what is really out there as you travel through the deepest regions of space? What lies at the heart of the Milky Way galaxy and at the edge of the known universe, where time began?

Astronomers have pointed the 27 giant dish antennas of the Very Large Array (VLA) radio telescope, on the Plains of San Agustin, New Mexico, at the heavens to begin answering these questions. Nearby, in our solar system, the VLA telescope has shown ice on the burning-hot planet of Mercury. Penetrating the clouds of dust at the center of our galaxy, the VLA has revealed in stunning detail the dense point of a black hole, a bizarre object of crushed mass and gravity, where light and matter swirl in and disappear for good. The VLA has also discovered a burst of brand-new star formation, future suns for their own worlds, in the spectacular Orion nebula, and it has probed "doughnuts" of gas flying out into space, the speeding farewell of a dying, exploding star.

Metric versus English Units		
Metric Unit	Abbreviation	Equals
centimeter	cm	0.4 inch
kilogram	kg	2.2 pounds
kilometer	km	0.6 mile
meter	m	3.3 feet
millimeter	mm	0.04 inch

The VLA antennas line up to form a Y.

An instrument that can reveal such wonders in the universe, the VLA has a strange beauty of its own. The VLA telescope's 27 massive white radio receivers are lined up in a Y shape, pointed at the sky. Built in 1980 and updated to use the most sophisticated computer and mechanical technology on Earth, the VLA is well equipped to hunt for strange objects and solve astronomical mysteries.

The VLA operates by receiving faint radio signals from outer space. Most of the signals are so faint, a blastingly strong signal for the VLA would be a cell phone going off on the moon, which is 384,400 kilometers (238,855 miles) away. To receive such weak signals, the signal-gathering dishes of the VLA must be huge.

A VLA antenna is moved along railroad tracks to a new position.
The orange machine is a 90-ton diesel-hydraulic transporter vehicle
especially designed for the VLA.

Each antenna in the VLA weighs 230 tons, and the dish is 25 meters
(82.5 feet) across. The antennas can be moved along railroad tracks to
separate them so that the resolution, or detail, of the images is better (the
maximum separation is 36 kilometers, or 21.6 miles). The antennas are set
in three 20.8-kilometer-long (12.5-mile-long) arms, and the antennas can
be spread out or bunched in four different ways, called the A, B, C, and D
configurations. At their widest, when the antennas are set in the A
configuration, they cover an area greater than that of Washington, D.C.
The antennas are set close together to improve sensitivity, or signal-
gathering power. To move just one arm of the giant antennas takes an
entire week of careful maneuvering.

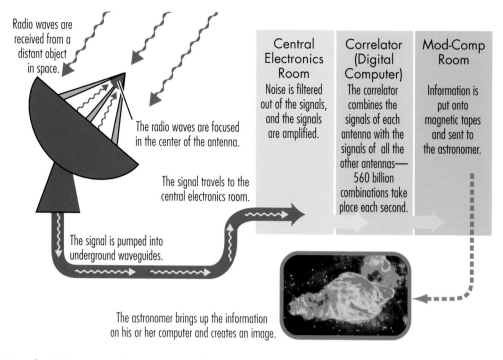

Central Electronics Room	Correlator (Digital Computer)	Mod-Comp Room
Noise is filtered out of the signals, and the signals are amplified.	The correlator combines the signals of each antenna with the signals of all the other antennas— 560 billion combinations take place each second.	Information is put onto magnetic tapes and sent to the astronomer.

Radio waves are received from a distant object in space.

The radio waves are focused in the center of the antenna.

The signal travels to the central electronics room.

The signal is pumped into underground waveguides.

The astronomer brings up the information on his or her computer and creates an image.

How the VLA turns radio wave signals into images. Radio waves from space are gathered at each dish antenna and focused at the center. Then a waveguide (a copper-lined steel pipe) from each antenna carries the signal to the central electronics room, where noise is filtered out and the faint signal is amplified. The signal then travels to the correlator, a digital computer, which combines the signals from all 27 antennas. Finally the information is put on magnetic tapes and sent to the astronomer. The astronomer creates an image of the object on a computer, assigning colors to the image based on the strength of the radio signal.

The signals that each dish picks up from space are collected in a box at the center of the dish. To hold down random noise, which results when the atoms that make up substances jiggle, this box must be kept very cold, at about 15 degrees Kelvin. That temperature is 464 degrees Fahrenheit below the freezing temperature of water and just 15 degrees above the coldest temperature possible, absolute zero, when even atoms stop moving. The signals run along a cable called a waveguide from each antenna to the VLA headquarters, a building at the telescope site, and are combined there by computer. A special-purpose computer called a correlator does the initial processing of the signals, performing 560 billion combinations each second. Then the information is transferred to magnetic tapes and can be sent to the astronomer who ordered the particular study. Although astronomers sign up to use the VLA for specific projects, operators at the VLA do the actual adjustment of the antennas.

Train wreck at the galactic center. Two galaxies are smashing into each other in this image to form a single galaxy, NGC 520. The green color is optical (regular) light, yellow is warm gas, and blue is cold (hydrogen) gas. The image is a composite made using the VLA and Kitt Peak National Observatory telescopes.

Because the signals from so many antennas, so far apart, are combined, the 27 antennas of the VLA are like one huge telescope miles across. When strong winds blow across the plains, the antennas can be moved into the fully upright position (the position of a cereal bowl) so that they catch less wind and don't tip over. At any time, one of the antennas is undergoing maintenance in a building on the VLA site—cleaning, painting, and repair or replacement of worn parts. (The paint used on the VLA is a special mix that costs hundreds of dollars a gallon.) Most of the electronic parts used in the VLA are made in the laboratories of the National Radio Astronomy Observatory (NRAO), a government agency that has branches and telescopes in many states. If you go into the basement at the NRAO building in Socorro, New Mexico, 60 miles from the VLA site, you will see technicians at work, building parts for the telescope.

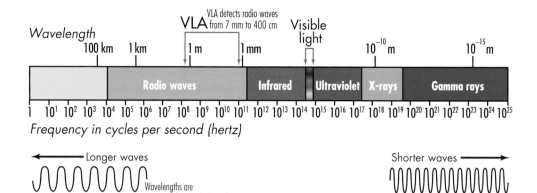

The electromagnetic spectrum. Electromagnetic waves with short wavelengths are high energy, like X-rays and gamma rays. Visible light rays and radio waves have longer wavelengths and are lower energy. The VLA detects radio waves at wavelengths of from 7 millimeters (0.07 of a meter, or about 0.28 inch) to 400 centimeters (about 13 feet). The frequency of electromagnetic waves is the number of waves that pass a point over a certain length of time, usually a second. The frequency can be given in cycles per second.

Both a radio telescope like the VLA and a regular AM radio receive radio waves: the VLA from outer space and your radio from a transmitter closer by. You might think that the radio waves the VLA and household radios pick up are noises, like bursts of static or the music a DJ spins from a city. In fact, radio waves make no more noise than visible light does. In ordinary radios, a receiver picks up radio waves sent by a radio broadcast tower. By spinning the radio's dial, you select the frequency of the signal (the radio station). The speakers convert the signal into vibrations—the sounds of music, news, or commercials.

Each of the 27 antennas of the VLA picks up radio waves. Because Earth rotates, the path from a radio source in the sky to each antenna is slightly different, which makes the radio waves each antenna receives slightly different. Sophisticated mathematics and technology transform these differences in the radio waves into images.

You may be familiar with optical, or light, telescopes—telescopes that magnify faraway objects and show them to you close up using regular visual light. Binoculars are this kind of telescope, and so is the Hubble Space Telescope, in orbit around Earth since 1990. Radio telescopes and optical telescopes both pick up *electromagnetic waves.*

The light that hits your eyes and enables you to see the blackboard, your friends, and other objects is in the form of electromagnetic waves. The *frequency* of electromagnetic waves is how many waves pass a point over a

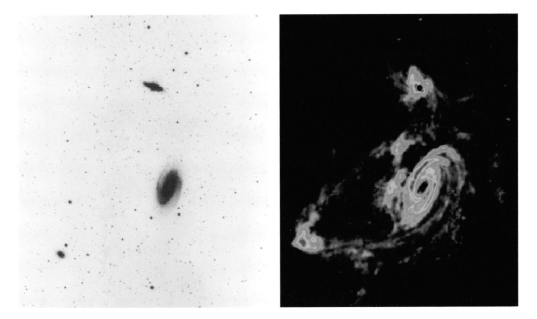

Two images of the radio (radio wave-emitting) galaxy M81, a spiral galaxy about 11 million light-years from Earth, measuring 50,000 light-years across. (Our own galaxy, the Milky Way galaxy, is a spiral similar to this one.) The image at the left was made with an optical telescope and shows mostly starlight. The image at the right, a radio image, reveals hydrogen gas. The radio image shows that M81 is not just one galaxy but a group of colliding galaxies. The radio image was made with the VLA in three of its four possible configurations and took more than 60 hours of observing time.

certain time. The *wavelength* is the distance from peak to peak in the waves. Light waves and radio waves are all part of the *electromagnetic spectrum*, but visible light, the light we see, is a very small part of it. Bumblebees can see ultraviolet light, which has a shorter wavelength than visible light, and some other insects can see infrared, which has a longer wavelength, as do radio waves. Radio telescopes can "see" radio waves. Radio and optical telescopes reveal very different aspects of an object in space.

Electromagnetic waves, both light and radio, move incredibly fast through space: at 300,000 kilometers per second (186,000 miles per second). That means light or radio waves travel to the moon in just 1.3 seconds! Light travels about 9.6 trillion kilometers (96 followed by 11 zeros, or 9,600,000,000,000 kilometers) (6,000,000,000,000 miles) in a year and would take about four years to reach the star nearest to Earth, Proxima Centauri. Because the distances between stars and galaxies are so huge, astronomers measure these distances in *light-years*. A light-year is equal to the distance light travels in one year.

Radio waves weren't discovered until the nineteenth century. In 1887 Heinrich Hertz, a German scientist, generated and detected radio waves by passing a spark between two electrodes. In the following decades, devices like home radios were invented to harness the new discovery. But no one thought of looking for radio waves in outer space.

Radio astronomy began by accident. In 1931, Karl Jansky was hired by AT&T to search for sources of static that might interfere with the use of radio waves for transatlantic communications. Jansky identified static from nearby and distant thunderstorms and random radio noise from devices on Earth. But that still left an unknown source of radio noise coming from a specific point in the sky. What Jansky had found was a radio hiss from the Milky Way galaxy. In 1939 his discovery was confirmed by an amateur astronomer named Grote Reber living in Wheaton, Illinois. Reber had built his own large radio telescope in his backyard, to the astonishment of his neighbors. But neither Jansky nor Reber could determine what was generating these cosmic radio waves.

After World War II, astronomers constructed more radio telescopes, with greater resolving power, or ability to see finer details, and greater sensitivity to the faint radio signals from space. In the 1970s, the NRAO built the VLA at a cost of over $78.6 million. In 1983 a team of astronomers led by Ron Ekers trained the VLA telescope on the exact center of our galaxy, on Jansky's mysterious radio source. This source had been named Sgr A* since it is found in the direction of the constellation Sagittarius. But the strange thing is, the source of the radio waves is tiny in astronomical terms—measuring only about a twentieth of the distance between the sun and Earth, or 7.44 million kilometers (4.46 million miles). That may sound like a huge distance, but the Milky Way galaxy is 8.6×10^{17} kilometers across (86 followed by 16 zeros, or 860,000,000,000,000,000) (5.4×10^{17} miles across). The radio object at our galaxy's center hardly moves, hanging almost motionless in the endless black of outer space, shrouded in the dust from millions of long-exploded stars.

But for the first time, using the VLA, astronomers saw Sgr A* for what it really was. At the very center of our galaxy is a monstrous black hole, with a mass of nearly 4 million suns, as relentlessly as a bathtub drain sucking in countless tons of matter, never to be seen in this universe again.

Grote Reber's backyard radio telescope in Wheaton, Illinois, in about 1938.

Other Famous Telescopes

The Very Large Array, a radio telescope, has made dramatic discoveries to advance our understanding of the universe. The VLA is a modern telescope, but the use of telescopes to scan the sky has a long history, dating four centuries back to Italian astronomer Galileo Galilei. Since Galileo's discoveries, telescopes have come a long way. They're more powerful, and different kinds have been invented. Other telescopes include the Hale Telescope, the Hubble Space Telescope, and the Chandra X-Ray Observatory.

Galileo didn't invent the telescope, but he was one of the first people to study the universe with one, in the year 1610. With his ordinary light telescope, which magnified objects just 20 times their real size, Galileo saw craters on the moon, discovered four of Jupiter's moons, and examined the rings of Saturn. Galileo's observations convinced him that Earth orbits the sun, instead of the sun orbiting Earth. At the time, the Catholic Church took the view that Earth was the center of the universe, and so the sun must orbit Earth. The church sentenced Galileo to life in prison for his heresy, or for disagreeing with the church.

Hale Telescope

A marvel of the twentieth century, the Hale Telescope began to operate in 1948 on top of Palomar Mountain in southern California. The telescope is named for astronomer George Ellery Hale, who designed it.

The Hale Telescope is a reflector telescope, which uses a curved mirror to gather and focus light rays so that an astronomer can see an image of an object in space. The telescope's mirror is coated with aluminum, a shiny surface that reflects light rays. The bigger the mirror, the more light rays it can collect. The Hale Telescope's mirror is 200 inches wide and weighs 14.5 tons.

The aluminum coating on the telescope's sensitive mirror must be replaced every two years. Workers first soap the mirror off with high-quality sea sponges. Then they wash the mirror with acid to remove the old surface. After many hours of washing, rewashing, and drying, the mirror is sealed with a new, thin coat of aluminum, sprayed on with a device that looks like a perfume atomizer.

Discoveries made with the Hale Telescope showed that the mysterious, brilliant objects called quasars are very powerful but far away and supported the theory that the universe is expanding. But the Hale has limits to what it can see. Any telescope on Earth must look through Earth's thick atmosphere. The atmosphere absorbs a lot of light. It also distorts light, which is what makes the stars seem to twinkle from Earth. Another problem for an Earth-based optical telescope is light pollution. When the Hale went into operation in 1948, not so many people lived in southern California, and it was a lot darker at night than it is now. Light from the growth of cities prevents the Hale Telescope from seeing the stars as well as it could.

When the Hubble Space Telescope went into orbit around Earth in 1990, it was free of Earth's atmosphere and light pollution from cities. The Hale's glory days seemed over.

But don't count the Hale out yet. Since 1948, the telescope has been upgraded with sensitive position sensors and high-speed computers. New electronic devices can detect very faint light from faraway objects. The telescope can now also detect infrared light, which is at longer wavelengths than visible light. The Hale Telescope is often used with other telescopes these days to form a complete picture of planets, stars, and galaxies.

In 1997 the Hale found two tiny new moons of Uranus, named Caliban and Sycorax. Caliban is only 80 kilometers (48 miles) in diameter and Sycorax is only 160 kilometers (96 miles) in diameter. The telescope was also part of a team, including the VLA and a NASA satellite, that recently studied a gamma ray burst 5 billion light-years from Earth.

Hubble Space Telescope

Launched in 1990 from the space shuttle Columbia, the Hubble Space Telescope is named for Edwin Hubble, who showed in the 1920s that other galaxies are very far away in space and that the universe is expanding. Orbiting 600 kilometers (360 miles) above Earth, the Hubble telescope is far above Earth's atmosphere and bright lights. The Hubble telescope, like the Hale, is an optical telescope and sees regular light waves, using mirrors to focus and magnify images, and can also detect infrared and ultraviolet light.

Glistening silver against blue-and-white Earth, the Hubble Space Telescope sails through space just days after the Columbia space shuttle released it into orbit. The rectangular "wings" are solar panels, which convert energy from the sun into electricity to power the telescope's scientific instruments, computers, and radio transmitters. The black dish on the front is a communications antenna so that the operations team on the ground can signal the telescope what to do. The two main computers are inside the telescope, right behind the solar panels. One computer gives orders to the scientific instruments on board, and the other controls systems like the gyroscope, which corrects the telescope's position in space. To the right of the telescope, an astronaut is at work.

In 1993, two and a half years after Hubble went into orbit, astronauts space-walked from the space shuttle Endeavor over to the telescope to correct the shape of its lens—it was too flat by two millionths of a meter, about 1/50 the thickness of a human hair. Since then, astronauts have visited Hubble many times to replace parts or install upgrades. The Hubble's design is modular: astronauts can easily remove just one piece and put in a new one.

The Hubble telescope has taken sensational pictures. Astronomers have turned the telescope on more than 25,000 objects in space. Among many other discoveries, Hubble has found that even small areas of the night sky are filled with galaxies and that some dusty regions in space may be like our own solar system was long ago.

In January 2004, NASA (the National Aeronautics and Space Administration) announced that the Hubble telescope will be allowed to degrade until it burns up in Earth's atmosphere, although NASA is reviewing this decision. NASA will shift its focus to manned moon and Mars visits, with astronauts expected to arrive on the moon as early as 2015 and on Mars sometime after 2030.

Chandra X-Ray Observatory

The Chandra X-Ray Observatory went into orbit around Earth in 1999. As you would expect from its name, the Chandra telescope detects X-rays, which have much shorter wavelengths and higher energy than visible light. The other part of the telescope's name comes from the astronomer Subrahmanyan Chandrasekhar, who made important discoveries about black holes. *Chandra* means "moon" or "luminous" in the language Sanskrit.

The space shuttle Columbia, ready for takeoff at the Kennedy Space Center in Florida. The Chandra X-Ray Observatory is in the tall metal box at the bottom left of the photo, about to be loaded onto the shuttle. Columbia carried the Chandra telescope into space and released it into orbit around Earth in 1999. Tragically, on February 1, 2003, the hardworking space shuttle broke up and was lost over the southern United States as it attempted to land.

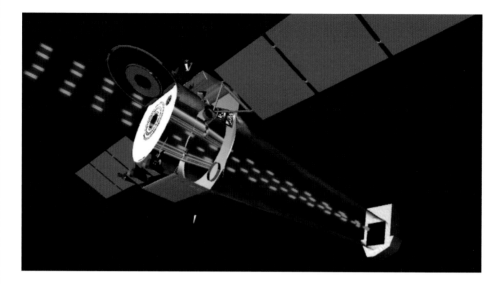

This drawing of the Chandra X-Ray Observatory shows X-rays (orange) being focused by the telescope's mirrors, then guided to the electronic detector so that an image can be made.

Astronomers had to put a telescope into orbit to detect X-rays from outer space because X-rays don't penetrate Earth's atmosphere. Life on Earth lucked out, because if the high-energy X-rays could reach Earth's surface, they'd zap living things to death. Astronomers didn't completely luck out, because a lot of objects in space give off X-rays.

Like the Hubble telescope, the Chandra Observatory works by reflecting electromagnetic waves with very precise mirrors. Hubble reflects visible light waves, and Chandra reflects X-rays. Because X-rays are so energetic, they hit Chandra's four pairs of barrel-shaped mirrors at an angle, like a pebble skipping across a pond. If the X-rays slammed into the mirrors head-on, they'd dig into them like a bullet. The reflected X-rays are narrowed onto an electronic detector to make an image of an object in space.

Black holes emit a lot of X-rays as they drag in pieces of their blown-up parent star and other material. The Chandra telescope has found many new black holes, including some medium-size ones, a new class of black holes. Chandra also showed that there are twice as many supermassive black holes as astronomers previously thought and detected eerie X-ray winds blowing from some of those big black holes.

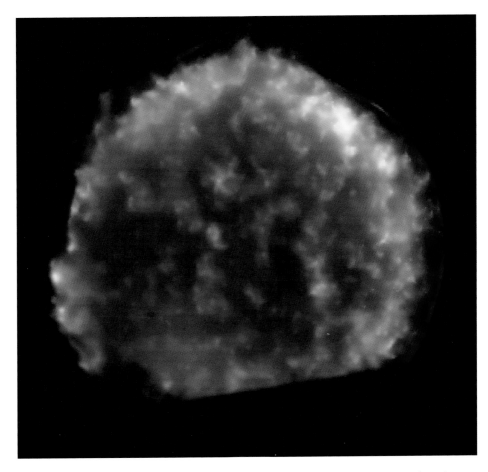

Here is Tycho's supernova remnant, the same one shown on page 44—but this image was taken with the Chandra X-Ray Observatory Center's X-ray telescope instead of with a radio telescope. In the image, the red shows low X-ray energies; the green, medium X-ray energies; and the blue, high X-ray energies. The remains of the exploded star are at a temperature of about 10 million degrees Celsius (18 million degrees Fahrenheit) and appear as splotches of yellow, green, and red. The blue at the outer edge is gas at a temperature of about 20 million degrees Celsius (36 million degrees Fahrenheit).

Radar image of Mercury. The dish antenna at Goldstone, California, used a powerful transmitter to bounce microwaves off the planet's surface; the signal was then collected by the VLA and a radar image formed. Red is a strong reflection of the signal, and yellow, green, and blue are weaker reflections (blue is weakest). The red reflection at the top of the planet, Mercury's north pole, is similar to the strong radar echo seen from the icy polar caps of Mars and the cold moons of Jupiter.

TWO

Probing the Planets

Red-Hot, Icy Mercury,
Killer Winds on Jupiter,
and Summering on Uranus

Scientists have used the VLA telescope to examine black holes and other extremely faraway objects in space, but they have also turned the VLA on the planets in our solar system, which are much closer to home. For centuries people have studied the planets in our solar system, first with ground-based telescopes, then with planetary missions and orbiting telescopes. The VLA, through its collection of radio signals, has provided new insights on the makeup of our nearest astronomical neighbors.

Earth, the eight other planets—Mercury, Venus, Mars, Jupiter, Saturn, Uranus, Neptune, and Pluto—and comets, asteroids, and moons make up the unusual cast of characters that is our solar system. All nine planets orbit the sun. Mercury, Venus, and Mars are the planets closest to Earth. The planets fall into two basic categories: mostly solid and relatively small—Mercury, Venus, Earth, Mars, and Pluto—or mostly gaseous and huge—Jupiter, Saturn, Uranus, and Neptune.

Our familiar solar system still holds great surprises. Astronomers using the VLA telescope found one such surprise on Mercury.

Planetary Facts

	Rotation Period*	Year Length**	Diameter	Surface Temperature	Average Distance from Sun	Moons
Mercury	58.7 days	0.2 yrs.	4,879 km (3,032 miles)	−173 to 427°C (−279 to 801°F)	57,909,175 km (35,983,095 miles)	0
Venus	243.0 days	0.6 yrs.	12,104 km (7,521 miles)	462°C (864°F)	108,208,930 km (67,237,910 miles)	0
Earth	23.9 hours	1.0 yr.	12,756 km (7,926 miles)	−88 to 58°C (−126 to 136°F)	149,597,890 km (92,955,820 miles)	1
Mars	24.6 hours	1.9 yrs.	6,794 km (4,222 miles)	−87 to −5°C (−125 to 23°F)	227,936,640 km (141,633,260 miles)	2
Jupiter	9.9 hours	11.9 yrs.	142,984 km (88,846 miles)	−148°C (−234°F)	778,412,020 km (483,682,810 miles)	63
Saturn	10.7 hours	29.4 yrs.	120,536 km (74,898 miles)	−178°C (−288°F)	1,426,725,400 km (885,904,700 miles)	46
Uranus	17.2 hours	84.0 yrs.	51,118 km (31,764 miles)	−216°C (−357°F)	2,870,972,200 km (1,783,939,400 miles)	27
Neptune	16.1 hours	164.8 yrs.	49,528 km (30,776 miles)	−214°C (−353°F)	4,498,252,900 km (2,795,084,800 miles)	13
Pluto	6.4 days	247.9 yrs.	2,302 km (1,430 miles)	−233 to −223°C (−387 to −369°F)	5,906,380,000 km (3,670,050,000 miles)	1

*Rotation period is measured in Earth days, hours, and minutes
**Year length is measured in Earth years (yrs.)

The solar system, from closest to the sun to farthest—Mercury, Venus, Earth, Mars, Jupiter, Saturn, Uranus, Neptune, and Pluto—showing the sizes of the planets compared to one another. Pluto's big moon, almost half the size of Pluto, is also shown. The distances from the planets to the sun aren't accurate in the figure—that would take too much room to show. The images of the planets come from the Voyager and Mariner 10 spacecraft and the Hubble Space Telescope.

Mercury

Of all the planets in the solar system, you probably wouldn't expect to find ice on Mercury, the planet closest to the sun, with its savage daytime surface temperatures. But scientists using the VLA did just that.

Rocky, airless Mercury, named for the speedy messenger of the Roman gods, is just 57,909,175 kilometers (35,983,095 miles) away from the sun. Mercury swings around the sun in an eccentric orbit, sometimes approaching the sun as near as 46,000,000 kilometers (28,580,000 miles) and sometimes moving as far away as 69,820,000 kilometers (43,380,000 miles). The time it takes Mercury to travel once around the sun, the Mercurian year, is just 88 Earth days, compared to 365 days in the Earth year. The long day on Mercury, though, lasts for almost 59 Earth days, since Mercury turns slowly on its axis. During that day the huge, blazing nearby sun fills the black sky, appearing almost three times as large as it does from Earth. No matter where Mercury is in its orbit, the side of the little planet that faces the sun bakes, reaching temperatures of 427 degrees Celsius (801 degrees Fahrenheit).

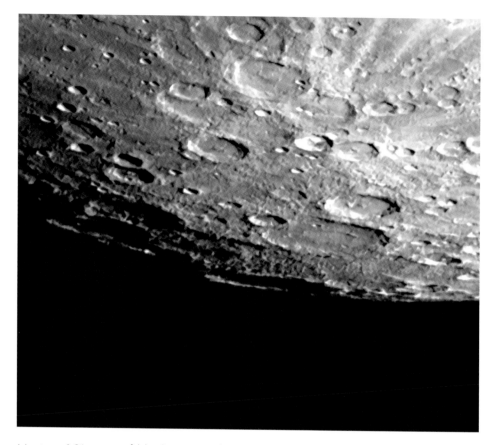

Mariner 10's view of bleak, cratered Mercury.

Mercury is a dead world—no volcanos spew lava across its surface, creating islands or mountains, and no earthquakes shake it, as they do our living Earth. Almost all of any atmosphere Mercury might have had long since boiled off in the heat of the sun, like water boiling away on a stove. Because Mercury is so small (Mercury is only about a third the size of Earth), its gravity isn't strong enough to hold the atoms of an atmosphere to the surface for long. The sun bombards Mercury with a solar wind of particles, pounding out a few atoms from Mercury's surface to make a thin, temporary atmosphere of mostly sodium, one of the two elements that make up table salt. Because Mercury has no atmosphere to hold in heat, the side away from the sun drops to a frigid minus 173 degrees Celsius (minus 288 degrees Fahrenheit) in the dark and silence of Mercury's midnight.

Earth's moon looks much like Mercury. Both are heavily cratered by rocks hitting the surface at high speed from outer space. Bright rays shoot out from craters on the moon and Mercury from these explosions. But unlike the moon, Mercury has long scarps, or low mountains with ridges—the longest scarp on Mercury is more than 500 kilometers long (300 miles long) and 1 kilometer high (0.6 mile high). Mercury's surface is wrinkled like a baked apple, and for the same reason: the planet was very hot when it first formed, probably out of hot chunks of rock colliding and melting together, then shrank as it cooled.

Mercury hasn't been studied as much as some of the other planets in the solar system, like Mars, first on the list for humans to visit, or beautiful Saturn, with its eye-grabbing ring system. Mercury is often hard to see with the naked eye or with optical telescopes since it's so close to the sun—the sun's glare in the Earth's sky drowns Mercury out. (Mercury can occasionally be glimpsed from Earth at twilight.) In 1974 and 1975, the Mariner 10 spacecraft swung around Venus, the heavily clouded planet between Mercury and Earth, and passed Mercury three times, photographing 45 percent of the planet's surface. Mariner 10 came closest to Mercury's surface, within 703 kilometers (437 miles), on March 29, 1974. The rest of Mercury remained a mystery, and no spacecraft ever returned.

In 1991 planetary scientists Duane Muhleman and Bryan Butler at the California Institute of Technology and Martin Slade at NASA's Jet Propulsion Laboratory used a half-million-watt transmitter (ordinary light-bulbs shine with about 75 watts) combined with the 70-meter (231-foot) dish antenna at Goldstone, California, to beam microwaves at Mercury. The waves bounced off the side of Mercury that had not been photographed by Mariner 10 and were collected at the VLA to form a radar image of the planet.

The radar image showed a strong reflection of the signal, indicated by red, at Mercury's north pole. The signal was similar to signals from the polar caps of Mars and the icy moons of Jupiter. The ice at Mercury's pole is not only frozen but very cold—it may be as cold as minus 148 degrees Celsius (minus 234 degrees Fahrenheit). Scientists know this because normal water ice, at the temperatures found on Earth, actually absorbs radio waves instead of reflecting them. The very cold ice can exist at the bottom of craters on Mercury because it is permanently shaded from the sun's glare. In 1994, the same team of scientists found water ice on Mercury's south pole.

As was done in the Mercury work, VLA images and information are often combined with those from other telescopes to round out the picture of an object in space. The Deep Space Network (DSN) is operated by the Jet Propulsion Laboratory for NASA and consists of three antennas placed 120 degrees of longitude apart (a circle is made up of 360 degrees) around the world: at Goldstone in California; near Madrid, Spain; and near Canberra, Australia. Because the antennas are spaced around the world, they can be in constant contact with spacecraft exploring the solar system. Signals from the DSN antennas control the spacecraft, and the antennas transmit the images and information the spacecraft collect.

Plans are under way for another spacecraft to visit Mercury. The spacecraft Messenger (short for Mercury Surface, Space Environment, Geochemistry, and Ranging) should reach orbit around Mercury by 2011. Messenger will examine Mercury's crust, volcanic history, and thin atmosphere—and take another look at the polar caps.

Jupiter

Jupiter is king of the planets in the solar system—more than 1,000 Earths would fit inside it. This giant planet is also renowned for its storms and winds. Finding out just how fast those winds are blasting around Jupiter has been a job for the VLA telescope, teaming this time with the Galileo spacecraft.

Jupiter is across from Mars, through the asteroid belt, which is a swarming minefield of different-size boulders. As the largest planet in the solar system, Jupiter is very bright in Earth's night sky, looking like a big star. People have seen Jupiter since ancient times without telescopes.

Jupiter is the first planet out from the sun in the group of planets called the Jovian planets—Jupiter, Saturn, Uranus, and Neptune. They are grouped because they are all very large and mostly made out of the gases hydrogen and helium. Jupiter is 142,984 kilometers in diameter (88,846 miles across), compared to Earth's 12,756 kilometers in diameter (7,926 miles across). The Jovian planets all have rings (although Saturn's rings are the most spectacular) and many moons, large and small. Ganymede, Jupiter's largest moon, is three-quarters as big as Mars.

Unlike Mercury, Jupiter has a thick, stormy atmosphere, made up of chemicals like methane and ammonia. On Earth, these chemicals are found more often in swamps or floor cleaners than in the air—luckily,

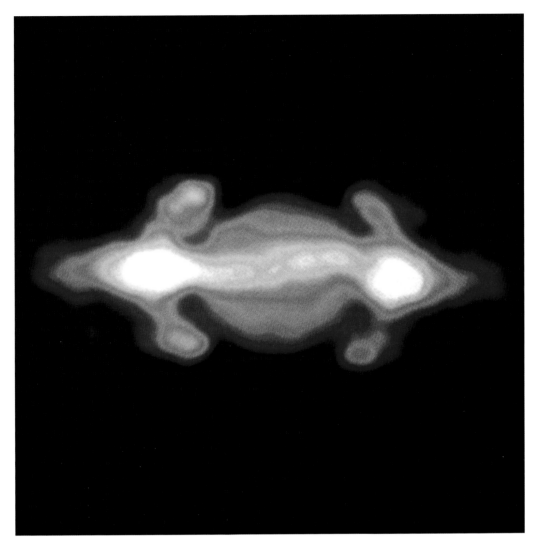

Radio Jupiter. Jupiter's strong magnetic field traps fast-moving electrons, which give off radio waves to produce the image seen here. The planet itself is at the middle of the image.

because Jupiter's atmosphere is completely unbreathable. Jupiter is so large, the atoms in its atmosphere are held to the planet by strong gravity and cannot escape. Lightning and thunder, many times more violent than on Earth, flash and boom in Jupiter's atmosphere. Jupiter's biggest storm, the Great Red Spot, is a hurricane about 14,000 kilometers wide (8,400 miles wide) and 40,000 kilometers long (24,000 miles long) that has lasted for hundreds of years.

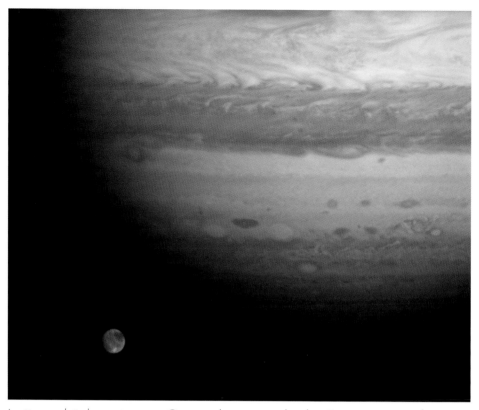

Jupiter and its largest moon, Ganymede, as seen by the Cassini spacecraft in December 2000. Ganymede is larger than either Mercury or Pluto. Cassini took this image 26.5 million kilometers (15.9 million miles) from Ganymede.

Almost all of Jupiter is liquid or gas, except for a tiny solid core. On a visit to Jupiter, you wouldn't want to stand on the core—it's thought to be extremely hot, about 40,000 degrees Celsius (72,000 degrees Fahrenheit). That heat is about all the warmth Jupiter has. The sun is 778 million kilometers away (467 million miles) and is just a small, brilliant ball in a nearly black sky. The cloud tops of Jupiter are cold—only about minus 153 degrees Celsius (minus 243 degrees Fahrenheit).

In 1995 astronomers teamed the VLA with the spacecraft Galileo to measure the wind speed deep in Jupiter's atmosphere. Launched in 1989, Galileo was the first spacecraft to enter Jupiter's atmosphere. Until 2003, Galileo orbited the planet, investigating Jupiter and its moons. In Galileo's work with the VLA, the spacecraft released a probe, hanging from a parachute, into Jupiter's atmosphere. The probe sank into the atmosphere, signaling as it went with a 25-watt transmitter. These very faint signals had to travel more than half a billion miles from Jupiter to the VLA on Earth and also to a similar radio telescope in Australia, the Australia Telescope Compact Array.

As Jupiter's winds battered the probe, they changed the frequency of the probe's radio wave signals. The VLA, with the help of powerful computers, detected these changes and could translate them into changes in wind speed. Deep in Jupiter's atmosphere, the probe and radio telescopes found winds blasting around the planet at 650 kilometers per hour (390 miles per hour). For the winds to be this deep, they must be powered by heat from Jupiter and not the sun, so they almost always blow this fast. The winds blow east *and* west. A typical day on Jupiter is something like a blustery March day on Earth—except Jupiter's day lasts only about nine hours and its winds are strong enough to slam you right around the planet.

Uranus

The gas giant Uranus is next out from the sun after Jupiter and Saturn. A very cold planet, with a swirling blue atmosphere made up mostly of methane gas, Uranus was not expected to have quick season changes like Earth from spring to summer, fall to winter. Observations made with the VLA, however, changed astronomers' expectations about the climate of faraway Uranus.

English astronomer William Herschel discovered Uranus in 1781 with an optical telescope—the planet is almost impossible to see from Earth without a telescope unless you know exactly where to look in the sky. Uranus is so distant from the sun (2,870,972,200 kilometers, or 1,783,939,400 billion miles), the planet takes 84 Earth years to make one orbit. The faint sun can barely heat Uranus, and Uranus's temperature is about minus 216 degrees Celsius (about minus 357 degrees Fahrenheit). The planet is made up mostly of water, methane, and ammonia and might have a small rocky core. We see Uranus as blue because the methane gas in the upper atmosphere absorbs red light, letting the blue light through. Like the other big, Jovian planets, Uranus has a whole flock of moons—27 have been discovered so far, five large and the rest small. The planet also has 11 rings, but they don't reflect light well, like a shiny stone does on Earth, and so the rings are dark.

The Hubble telescope's view of Uranus, showing its rings. This image is false color, unlike some of Hubble's images, which show objects in ordinary visible light. Blue and green are where sunlight can pierce deep through clear parts of Uranus's atmosphere. Orange and red show high clouds, like the high, wispy cirrus clouds on Earth. Uranus's cloud tops of methane make the planet's real color a mostly uniform blue.

No one has yet taken a summer vacation on Uranus, but the huge blue planet does have a summer. Earth has summer (and spring, fall, and winter) because our planet is tipped in space like a leaning, spinning top. At different times of the year, as Earth orbits the sun, more sunlight shines on the part of the planet that is tipped closest to the sun. Then it's summer on that half of Earth. When it's winter in Australia, the bottom half of Earth, or southern hemisphere, is tipped away from the sun. At the same time the top half of Earth, the northern hemisphere, is tipped toward the sun, and it's summer in the United States.

But Uranus's summer is much different from Earth's. Water stays rock-hard, permanently frozen during the planet's icy summer (you won't find any beaches on Uranus). Also, Uranus is tipped on its side. It rotates like a rolling penny (although it rolls in place) instead of a spinning top. No one knows why Uranus tipped over. Something monstrously huge may have slammed into it, like a massive asteroid. But because Uranus is on its side, first its north pole is closest to the sun, then its south pole, not the equator, which is always closest to the sun on Earth. Uranus's summer lasts as long as its pole points at the sun. Since a year on Uranus equals 84 Earth years, Uranus's summer lasts a fourth of that time, or 21 years. Spring, fall, and winter are each just as long.

Like Mercury, Uranus is an unpopular tourist destination. Only one unmanned spacecraft has ever visited Uranus—Voyager 2. Voyager 2 swung around Jupiter and Saturn first, studying those two planets, then arrived at Uranus in January 1986. Finally Voyager whizzed by Neptune, the next-to-last planet in the solar system, in August 1989. After studies of Neptune, Voyager 2 traveled into outer space, a messenger of Earth's civilization if anyone is out there to find it.

Astronomers working with the VLA studied changes in Uranus's atmosphere over the years 1981, 1985, 1989, and 2002. During these years, the season on Uranus changed from early summer to early fall. Surprisingly, the atmosphere changed rapidly over this time: the contrast between light and dark areas increased quite a bit. No one had expected that changes could occur so fast and go so deep on such a large planet. The light and dark areas probably indicate different amounts of ammonia and water. Through 2007, the astronomers plan to keep watch on Uranus to look for more changes in the atmosphere.

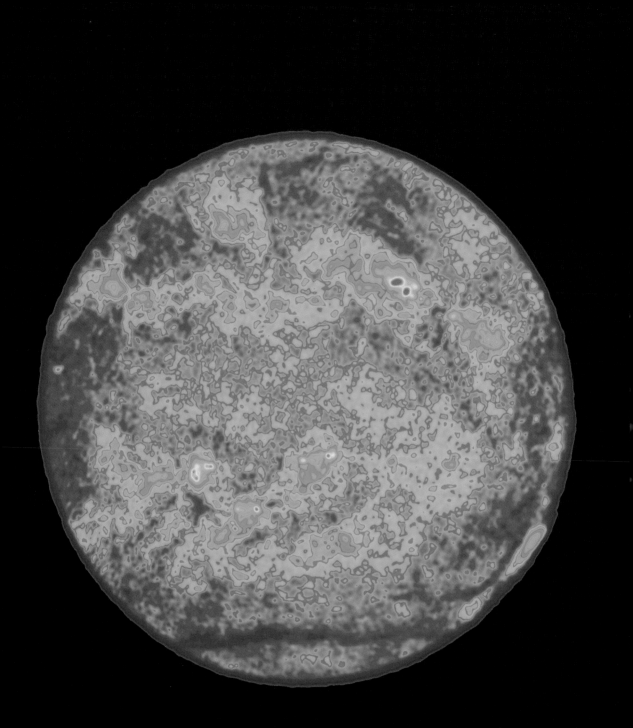

The sun, our solar system's star, is a strong source of radio waves. The red spots show the hottest temperatures, about 1 million degrees. On this particular day, sunspots could be seen in these areas. Yellow, green, and blue areas are cooler. The radio sun is about 20,000 kilometers (12,000 miles) bigger than the sun you see in the sky, or in visible light.

THREE

Doughnuts in Space
The Life and Explosive End of Stars Like the Sun

Astronomers working with the VLA radio telescope have made some splendid discoveries about the solar system, either just using the VLA telescope or teaming it with other telescopes and with spacecraft. These discoveries have helped us learn about planetary surfaces and atmospheres and have led to a greater understanding of how the solar system works. Also, if astronauts visit the planets someday, they will have a much better idea of what to expect, which could be crucial to staying alive.

Many of the VLA's most awesome discoveries, though, have been outside our solar system. After all, everyone knew Mercury was there before the VLA studied it. But no one knew what gamma ray bursters—death stars—were until the VLA provided an explanation. To understand these strange stars and the other odd paths the life of a star can take, we have to start with understanding our solar system's own star, the sun.

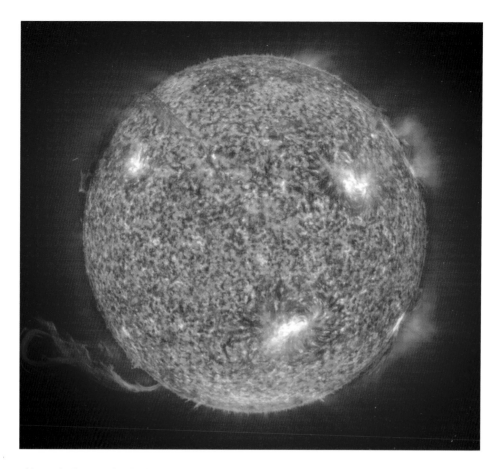

Although the sun looks like a smooth yellow ball from Earth, it is in fact a turbulent, hot sea of gas. In this image taken by the Extreme Ultraviolet Imaging Telescope (EIT), a gigantic solar flare erupts from the photosphere of the sun, arcing into space.

The Sun: An Ordinary Star

At the center of our solar system, surrounded by the nine planets, is the sun. The sun is an average-size yellow star, not old but not young—middle-aged. The sun is a bit unusual, since most nearby stars are smaller and cooler, and many of them are double: two or even three stars orbit each other. Of course, the sun is anything but ordinary to us. It's the source of all light and heat, making life possible on Earth. If the sun shut down now, soon Earth's surface would be frigid cold, pitch black, and dead, with deep frozen oceans of ice.

The Sun	
Layer	Temperature (Degrees Celsius)
Core	8 to 16 million degrees
Radiative and convective zones	8 million to 7,000 degrees
Photosphere	7,000 degrees
Chromosphere	7,000 to 1 million degrees
Corona	1 million to 2 million degrees

Luckily the sun is a stable, steadily burning star. But if you looked closely at the sun's surface, you might not expect the sun to shed warm, even light. Although from Earth the sun appears as a perfect round ball in the sky, it is actually a roiling, churning furnace. The sun is immense: 1,391,000 kilometers in diameter (864,400 miles across at its equator). It only looks the size of the moon because the moon is much closer. If Earth were held up right next to the sun, the sun would be 100 times bigger.

The sun has no solid surface. It can be divided into layers according to temperature and what is happening in each layer. At the center of the sun is the core, which is where the sun creates most of its energy. The core is very hot: about 8 million to 16 million degrees Celsius (when you convert this temperature to degrees Fahrenheit, the number is almost the same). Next out from the core are the *radiative* and *convective* zones, where the sun's temperature cools and the energy from the core worms its way outward in the form of *photons*, or particles of light energy. Light can be thought of as either electromagnetic waves or small packets of energy—the amount of energy is the same. Here it's simpler to look at the light energy as packets, traveling through the sun's layers and bumping into the sun's atoms, losing energy along the way.

The photons released from the core finally reach the surface, or *photosphere*, of the sun in about 200,000 years. By then the photons have lost enough energy to be visible radiation, or sunlight. In about eight minutes the sunlight zips through space to Earth.

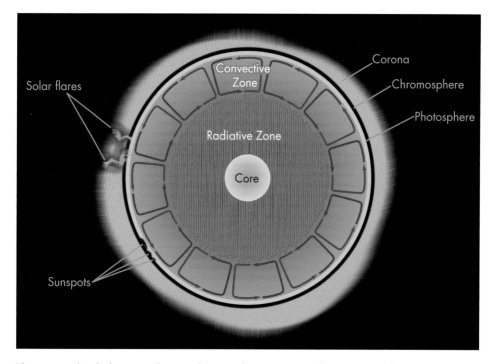

The sun is divided into six layers: the very hot core (8 million to 16 million degrees Celsius); the radiative zone and the convective zone, which move energy out from the core in the form of photons, or particles of light, and are cooler than the core; the sun's surface, or photosphere (about 7,000 degrees Celsius); the see-through chromosphere (with temperatures back up to 1 million degrees Celsius); and the intensely hot corona (topping out at about 2 million degrees Celsius).

The photosphere bubbles and cooks at about 7,000 degrees Celsius. As you go outward in the sun from the photosphere, the temperature quickly rises—strangely, the next layer out of the sun, the see-through *chromosphere*, is much hotter than the photosphere: up to 1 million degrees Celsius. The increase in temperature has to do with the sun's magnetic field. Magnetic fields store energy—you know this if you've ever tried to pull two magnets apart. Electric currents or particles moving through the magnetism in the chromosphere can create heat energy.

Solar flares arc off the photosphere, exploding into space. *Sunspots*, dark, cooler areas, move across the sun's surface in 11-year cycles. Solar flares and sunspots often occur together. They are caused by complicated flows and interactions of the sun's magnetic field and electric currents.

The *corona*, the outermost layer of the sun, is extremely hot—in some parts it tops out at 2 million degrees Celsius. It's probably so hot for the same reason the chromosphere is—electric currents or particles interact with magnetic fields. The corona is the brilliant halo of the sun that is seen during solar eclipses, when the sun is blacked out by the moon moving in front of it.

During the nineteenth century, scientists thought that the sun must burn coal to create its energy, but soon they had to throw out that idea. Evidence of the great age of some rocks on Earth proved that Earth, and so the sun, are very old. If the sun burned coal, it would have burned out in just a few thousands of years. Scientists had to look for a new explanation of how the sun created energy.

Albert Einstein, as he so often did, provided the answer. He proposed his famous equation $E = mc^2$, where E equals the amount of energy, m equals the mass of an object, and c^2 equals the speed of light, c, times itself, which, we saw, is a huge number—300,000 kilometers per second (186,000 miles per second). So even a small mass stores a lot of energy. It turned out that the sun transforms mass into energy. Nature can turn mass into energy in two ways: by *nuclear fission*, where atoms are split, and by *nuclear fusion*, where atoms fuse together. The sun creates energy through nuclear fusion. Basically, in the sun four hydrogen atoms are fused to create one atom of the element helium. In the process, a tiny amount of energy is released. But because the sun is so huge, all the fusion reactions add up to an incredible source of power. High pressure and temperature are needed to start and keep up the fusion reactions— the temperature must be at least 8 million degrees Celsius. That's why only the sun's core participates in fusion.

The sun has steadily released its energy for billions of years. It is made up of about 74 percent hydrogen and 25 percent helium (the remaining 1 percent is metals). But the sun can't burn all its hydrogen—only the hydrogen in or close to the core. That's only about 10 percent. Even so, the sun should burn steadily for another 5 billion years.

The VLA has participated in many studies of the sun, including teaming with an X-ray telescope to examine solar flares. But the VLA's work on stars outside our solar system has resulted in many strange new discoveries—not only about stars' life, but their birth and death.

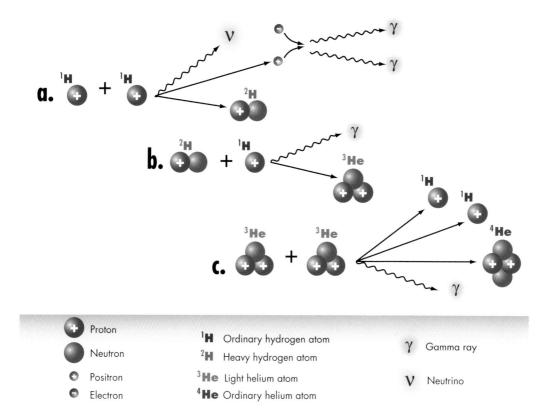

⊕ Proton	¹**H** Ordinary hydrogen atom	γ Gamma ray
⬤ Neutron	²**H** Heavy hydrogen atom	
⊕ Positron	³**He** Light helium atom	ν Neutrino
⊖ Electron	⁴**He** Ordinary helium atom	

Protons, neutrons, and electrons make up atoms. A proton has a positive electric charge, an electron has a negative charge, and a neutron doesn't have an electric charge, either positive or negative. Protons and electrons are like positive and negative numbers—they cancel each other out. An atom made up of one proton and one electron—which is a hydrogen atom—doesn't have a charge.

The process of nuclear fusion, the way the sun makes energy in its core, is shown above. First two regular hydrogen atoms join to form heavy hydrogen, ^2H (a). The squiggle labeled with the lowercase Greek letter ν (nu) is a neutrino—it is a neutral particle without mass or with very little mass that zips away when the two hydrogen atoms combine. An electron meets with its exact opposite, a positron, which is antimatter, and the electron and positron annihilate each other. Their mass turns into light energy: two very energetic photons called gamma rays fly off.

In the second step of nuclear fusion (b), the heavy hydrogen combines with another atom of regular hydrogen to make a light form of helium, which is missing a neutron, and also releases another gamma ray. In the last step of nuclear fusion (c), two light helium atoms combine to form one regular helium atom, releasing two regular hydrogen atoms and a gamma ray.

In this nuclear fusion reaction, a small amount of mass is converted into energy, following Einstein's equation $E = mc^2$. After 200,000 years, the gamma rays wind down in energy from colliding with other particles inside the sun and reach the sun's surface as visible light. About eight minutes later that visible light reaches Earth.

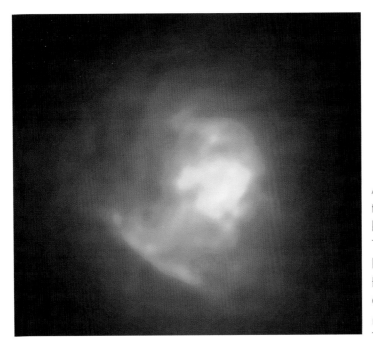

A giant cloud of gas, the Orion nebula is a birthplace for stars. This image was made by combining images from the VLA and the Greenbank single-dish radio telescope.

Star Nurseries

Like people, stars are born, live for a while, and die. Stars, though, are born in giant clouds of gas and dust. Gravity pulls the cloud together, and it heats up. When the temperature is high enough to start a fusion reaction, a star is born. For a star like our sun, the process from dust cloud to regular star is about 50 million years. The Milky Way galaxy, the big spiral-shaped collection of stars we live in, produces about 10 new stars a year.

Optical telescopes (and eyes) can't see through the dust into the star nurseries. But radio telescopes can—the radio waves they receive aren't blocked by dust. In the year 2000 astronomers trained the VLA on the Orion nebula, a beautiful giant cloud of gas about 1,500 light-years from Earth. The Orion nebula is full of young, hot stars, at temperatures of about 30,000 degrees Celsius. The VLA observations were combined with those from the Green Bank radio telescope, a single-dish telescope in West Virginia. The VLA and Green Bank telescopes were used together to get a better image of the nebula—single-dish radio telescopes like the Green Bank can detect large-scale structure well, but the many, spread-out antennas of the VLA are better at resolving detail. Combining the images from the two telescopes is difficult and requires fancy computer work. But the end result is a clearer image of the nebula—which is much needed, since the cloud's dust hides a good deal of what is going on inside.

Types of Stars

Class	Color	Surface Temperature (Degrees Celsius)
O	Bluish white	30,000
B	Bluish white	11,000 to 30,000
A	Bluish white	7,500 to 11,000
F	Bluish white to white	6,000 to 7,500
G	White to yellowish white	5,000 to 6,000
K	Yellowish orange	3,500 to 5,000
M	Reddish	3,500 and lower

The images from the two telescopes show ionized hydrogen gas, which has been stripped of electrons, flowing out from the young, hot stars in the center of the nebula. This gas hits colder gas at the edge of the nebula, forming a shock wave, which causes the gas to collapse and form stars.

The VLA has detected a pair of binary stars, each surrounded by its own dust disk. Planets are thought to often form from the dusty disks around new stars as the material in them clumps together. The VLA finding shows that not only single-star systems, like our sun, can form planets but also double-star systems. Since double stars are much commoner in the universe than single stars, a lot of planets may be out there. Perhaps some of them support life.

Giant dust clouds can turn into single stars, double stars, triple stars, or even more. The stars are of different sizes and temperatures. The biggest, hottest stars are bluish white in color and have surface temperatures of about 30,000 degrees Celsius. These are called *type O* stars. The coolest stars, *type M*, are just 3,500 degrees Celsius or less. Our sun, a *type G* star, is in the middle, with a surface temperature of about 5,500 degrees Celsius.

The Death of the Sun—a Planetary Nebula

The VLA has caught images of stars' deaths, including the first explosion and final spectacular grave marker splashed across the sky. One day, about 5 billion years in the future, the sun too will die. It will run out of gas—out of hydrogen to fuse into helium. Then humanity will have to leave Earth or face the catastrophic consequences.

Stars meet different fates. The more massive the star, the shorter its lifetime. The sun, a medium-size star, will live for about 10 billion years. Then it will run out of hydrogen in its core, and the nuclear reaction that fuses hydrogen to helium will stop, except in a shell around the core, where some hydrogen is still left.

Gravity will cause the sun's core to shrink. This heats it up, and it passes that heat along to the shell around the core, making the reactions go faster there and produce more energy. The sun will get brighter and its outer layers will blow out. As they blow out, they cool, and the sun's color will change from yellow to red. At this point it will seriously be time to leave Earth because the sun will become so huge, it will stretch out across Earth's orbit, rip off its atmosphere, and melt Earth's surface down 100 yards (90 meters). In less than 200 years after that, the sun will drag Earth into its core and vaporize it. At this stage the sun is a *red giant*.

But the sun's life won't be finished yet. Its core is now hot enough to start a new kind of nuclear reaction: fusing three heliums to one carbon atom. When the core runs out of helium, the same thing happens as when the core ran out of hydrogen: a ring around the core now fuses helium. The sun's core shrinks, the helium reactions in the ring go faster, and the sun's outer layers expand again. So the sun becomes a red giant twice.

The sun is now very unstable. Massive nuclear explosions shake it, and the sun brightens and dims by 50 percent in just a few years. A *superwind* blows out from the sun's surface, ripping off its outer layers. A shell of gas, or *planetary nebula*, forms around what is left of the sun, a hot core. Ultraviolet radiation from this hot core makes the shell of gas glow.

The first planetary nebula was sighted by French astronomer Charles Messier in 1764. Planetary nebulas have nothing to do with planets. The name comes from William Herschel, who discovered Uranus. While doing a survey of the sky, he found more objects like Messier's. Perhaps stuck on Uranus, Herschel decided that these new objects looked like Uranus through a telescope and called them planetary nebulas. The name also stuck. About

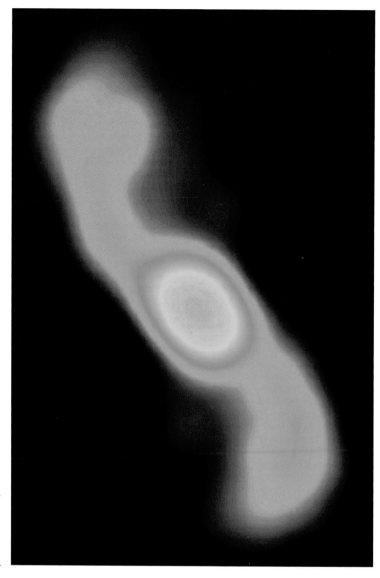

The planetary nebula K3-35 is 16,000 light-years from Earth. The VLA caught this old star during the brief time that it is transforming into a planetary nebula, an exploded star.

1,600 planetary nebulas have been found. Sometimes they are shaped like doughnuts, and sometimes they form weird, beautiful shapes.

The VLA has captured images of a star that is just starting to form a planetary nebula. This was a first, because the planetary nebula stage of a star's life doesn't last long. The nebula, called K3-35, is 16,000 light-years from Earth, in the constellation Vulpecula ("Small Fox"). This planetary nebula has the classic doughnut shape, with lobes of outflowing material continuing past the doughnut.

Astronomers using the VLA have detected radio waves emitted by water molecules in K3-35. Once a star is a full-blown planetary nebula, ultraviolet light, produced by what is left of the star as it heats up, would break up and so destroy the water molecules. Since water molecules are still around K3-35, the star must have just started down the path to becoming a planetary nebula—in about a hundred years, the water molecules would all be gone, zapped by the ultraviolet light. The star is thought to have begun its planetary nebula stage in 1984, not long ago at all. Best of all, astronomers can now keep watching K3-35 and see how events unfold.

After the hydrogen burning, the helium burning, the puffing up and down into a red giant, and the glorious burst into a planetary nebula, the sun will finally quiet down. The hot core left is now a *white dwarf* star, made mostly of carbon. For thousands of millions of years the small white dwarf will cool, until finally it is completely cold and dark. The sun, now a *black dwarf*, will be dead.

But the death of stars means life for other stars—and for us. The dying stars throw out dust and gas. Sometimes that dust and gas collect to make new stars, planets, and finally people. We aren't made mostly of hydrogen and helium, like the sun. As stars explode and die, hydrogen and helium fuse to make heavier elements, like carbon, oxygen, and iron. Those elements and many others make up the human body.

Astronomers have studied the red giant star Betelgeuse with the VLA to learn more about how dust and gas are thrown off old stars. Betelgeuse is about 430 light-years away, in the constellation Orion. This red giant is 600 times bigger than the sun. Betelgeuse is so big, if our solar system traded stars and Betelgeuse was put in the middle of the solar system instead of the sun, the bloated old star would swallow all the planets out to Mars and stretch into the asteroid belt.

Radio wave measurements made of Betelgeuse's atmosphere with the VLA show that huge plumes of gas chug up from the star's surface and fill the atmosphere. The VLA observations also showed that the atmosphere is cooler than expected close to the star's surface. That means dust grains can collect near the surface and then get a push into space from the burning star's radiation. If the dust grains formed farther away from the star, they wouldn't get a push. Red giant stars throw out huge amounts of gas and dust, and so knowing how they do it will help explain how the heavier elements make their way into new stars—and new life.

The new black hole or neutron star (shown in blue) at the center of just-discovered supernova 1986J. The other colors are the expanding shell of material thrown off the star in the supernova blast. This is an artist's impression.

FOUR

Death Stars and Starving Black Holes

The universe is full of extremely violent, often giant stars and remains of stars, with conditions very unlike those on Earth. The VLA telescope has helped to determine what these objects are and how they produce such violence. Neutron stars, which are the squashed remains of exploded massive stars, and black holes near and far, including Sgr A*, the black hole at the center of our Milky Way galaxy, have all been studied with the VLA. In a huge discovery, the VLA solved the origin of mysterious high-energy waves from space—the spewing of death stars.

Intense as the death of a sunlike star is, it's nothing compared to what happens when a more massive star dies. These stars live life in the fast lane: their greater temperatures and pressures cause them to burn up their hydrogen fuel much faster than a less massive, cooler star. Type O stars live just millions of years. When massive stars die, they explode into *supernovas*.

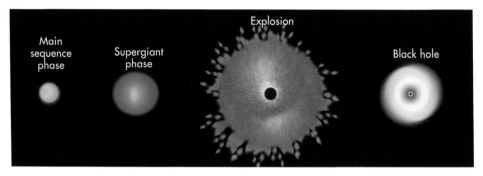

Stars go along for millions of years, steadily burning energy in the process of nuclear fusion. When a massive star's hydrogen supply runs low, it blows up, then shrinks, the same as smaller stars. But a massive star doesn't collapse into a white dwarf and then burn out into a dead cinder, like a small star. When the mass of a star is eight or more times that of the sun, the star can collapse into a neutron star or a black hole, its matter densely packed into a small space.

When a massive star runs out of hydrogen gas to burn, it fuses heavier elements, finally ending with iron, which is much heavier than hydrogen. But iron takes energy to make; the process doesn't release energy. So the fusion reactions stop. The star's core starts to collapse under the force of gravity, and the core whizzes inward at 70,000 kilometers per second (42,000 miles per second). The very heavy elements in the universe, like plutonium, are created in this process. Some of the core material bounces back, forming a shock wave that races out from the core and flings heavy elements into space.

What is left of the massive star has one of two possible fates: it can become a neutron star or a black hole.

Heavy Marshmallows

A *neutron star* is a strange object that forms when the gravity of a collapsing star presses electrons and protons together into neutrons. The force of gravity is so strong on a neutron star, a marshmallow dropped on one would explode with the energy of an atomic bomb. Neutron stars can form from stars that are between 1.3 and about three times the mass of the sun. The squashed, dense star is very small: its diameter is between 10 and 20 kilometers (six and 12 miles), depending on how big the star was to begin with.

A quickly spinning neutron star is called a *pulsar*. Pulsars whip around about once every millisecond (a thousandth of a second). As they spin, they

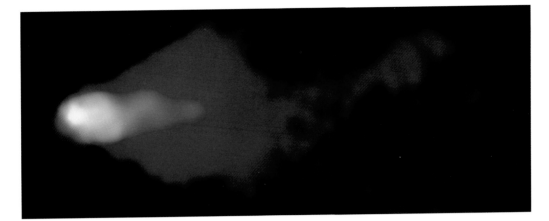

The Mouse, shown here in X-rays (gold) and radio waves (blue), is a pulsar that has a tail 55 light-years long. The pulsar, what is left after the explosion of a massive star, is speeding through space at 1.3 million miles per hour. The pulsar itself is the bright part of the image that forms the Mouse's head. The tail is formed as the rapidly spinning pulsar races through the interstellar gas.

shoot out pulses of electromagnetic waves, including radio waves, light waves, and X-rays. The pulses are quick and precise, lasting a few hundred milliseconds or less. Astronomers have found hundreds of pulsars.

Most supernovas took place a long time ago, before humans recorded history. But Chinese astronomers in the year 1054 observed a supernova explosion. They called it a guest star, and the remains of the supernova explosion became the Crab Nebula. It's 6,500 light-years from Earth, so the explosion actually took place a long time ago—the light had to travel all that distance before finally reaching Earth in 1054. In the Crab Nebula is a pulsar.

Since pulsars give off radio waves, the VLA can be used to study them. Recently astronomers used the VLA to examine the pulsar B1951+32, which is surrounded by the remains of the star's supernova explosion, called CTB 80. Both the pulsar and the shell are nearly 8,000 light-years from Earth. The pulsar is zooming out from the center of the shell at more than 800,000 kilometers per hour (more than 500,000 miles per hour). This was the first time that actual measurements showed a pulsar moving away from the supernova remnant.

The VLA astronomers measured B1951+32's position in 1989, 1991, 1993, and 2000. The Very Long Baseline Array, another radio telescope (more about that in chapter 6), also took measurements in the year 2000 to improve the study's precision. The combined measurements showed

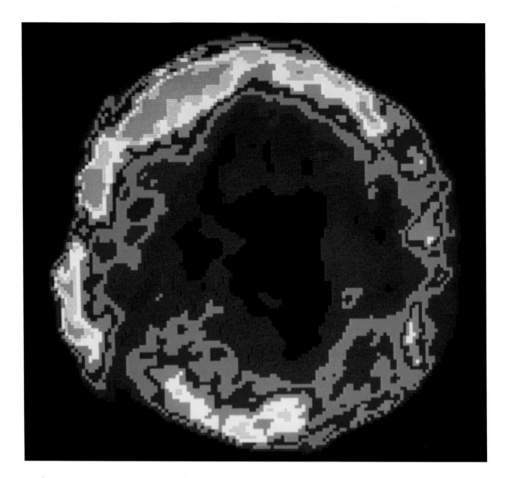

Tycho's supernova remnant, shown here in a VLA radio image, is left from a supernova explosion that took place in 1572. The supernova is named for Danish astronomer Tycho Brahe, who observed it at that time. Since then, the shell has expanded, growing bigger and bigger. The supernova remains are about 7,500 light-years from Earth.

that the pulsar took about 64,000 years to travel from the actual site of the supernova explosion to where it is now. So the pulsar is 64,000 years old.

Or is it?

Before the VLA work, astronomers had calculated the age of pulsars by measuring the rotation rate, or rate at which the pulsar spins, and then watching over time how much the rotation rate slowed. The pulsar slows as it loses energy. The old method gave an age for B1951+32 of 107,000 years, quite different from the VLA's number of 64,000 years.

Now astronomers will have to rethink their theories of how pulsars are born and what age all of them really are. In the end, all the new data and rethinking will lead to a better understanding of pulsars.

An artist's idea of what happens around a black hole. The black hole itself is invisible at the very center of the whirlpool. Two jets of energetic particles moving at high speeds beam out from the center of the hole. The whirlpool of circles is material drawn to the black hole by its powerful gravity and orbiting it.

No Escape—Swallowed by a Black Hole

When a star dies that is eight times or more as massive as the sun, its material just keeps collapsing under the force of gravity. The material is squeezed tighter and tighter until it vanishes from sight. What mass is left of the star is still there; it's just not taking up space. But the invisible star has a huge field of gravity that pulls on everything around it—including light. Even light can't escape because the force of gravity pulls it back down, which is why the leftover star can't be seen. The star is now a *black hole*.

The black hole pulls on material surrounding it, forming a ring of material called an *accretion disk*. If the black hole was part of a binary star pair when it was a star, the black hole can pull material off its companion star if it's close enough. That material gives off energy, which telescopes, including the VLA, can detect.

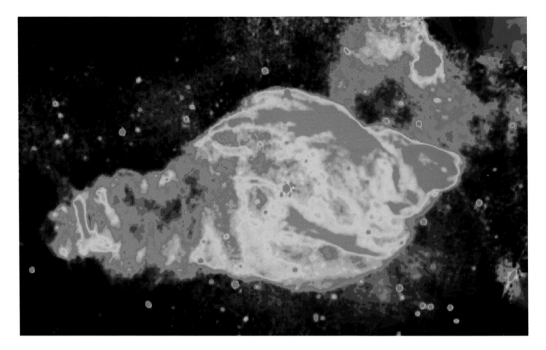

This VLA radio image shows the seashell-shaped supernova remnant W50, what is left after the explosion of a massive star. Jets of particles from the center are pushing the shell out. Red shows the brightest regions, yellow is less bright, green less bright still, and blue the least bright. The right side of the shell is brighter than the left because the explosion of the star pushed more gas there. The central object powering the action (the red dot at the center) is either a neutron star or a black hole. W50 is 10,000 to 16,000 light-years from Earth.

What would happen if you got near a black hole? If it had formed from a single, huge star with a mass about 10 times that of the sun, you would see a black patch of sky about 30 kilometers across (18 miles across). If you got within 3,000 kilometers (1,800 miles) of this black hole, the first thing that would happen is you'd be torn to pieces. But pretend that didn't happen. Then the black hole's gravity would just pull on you and stretch you out.

Suppose you were with a friend, and your friend (who also only stretched and didn't explode) jumped right into the black hole. He'd instantly be squeezed to zero volume—to a dot, then no dot at all. But you wouldn't see him disappear. The last thing you'd see was him falling toward the black hole across a boundary called the *event horizon*. The bigger the hole, the more gravity pull it has, and the farther away its event horizon is. Once your friend crosses the event horizon, light can't reach you anymore from the black hole, and you need to see light to know what's going on.

Not just light behaves bizarrely around a black hole—time slows down. If you were wearing a watch and your friend who jumped into the black hole was too, when you looked at his watch, it would show time going slower than on yours as he got closer to the black hole. Finally, at the event horizon, you would see that his watch had completely stopped. From your point of view, he would be frozen, stretched, falling into the black hole forever. From his point of view, his watch would keep going. He'd experience falling into the black hole and being ground to a point with infinite density—whatever *that* would feel like.

In 1986 the very youngest black hole yet was discovered in the galaxy NGC 891, about 30 million light-years from Earth. The supernova explosion that had left the black hole occurred about three years earlier. The VLA, working with the Very Long Baseline Array and other radio telescopes, made images showing details of the explosion. At the center is a bright object. The brightness is caused either by a black hole pulling in material or the action of a young pulsar, or neutron star. Astronomers plan to watch the new black hole or neutron star to see which it is and to learn more about the physics of these objects as they develop.

Gamma Ray Bursters

In July 1967, U.S. satellites were orbiting Earth, looking for violations of the 1963 Nuclear Test Ban treaty—signs that the Soviet Union was exploding nuclear weapons in Earth's atmosphere. High-ranking U.S. government officials panicked when the satellites detected energetic bursts of gamma rays. The bursts usually took place two or three times a day. Had the Soviet Union broken the treaty? That kind of conflict could lead to World War III.

But it turned out that the gamma rays were coming from all over the sky, and so they couldn't be the flash of a random nuclear weapon. A burst in January 1999 came from a source the size of the sun but had a brightness 10^{19} times greater than the sun's.

Gamma ray bursts (GRBs) are the most energetic explosions known in the universe: a burster known as GRB 970508 released in just 15 seconds almost 10 times as much energy as the sun will release over its entire 10-billion-year lifetime. Luckily this GRB was about 7 billion light-years away—if it had been even a few thousand light-years away, Earth would have been dosed with enough radiation to kill just about everything alive. (That's why GRBs got the name death stars.) Another GRB outshone the

What the gamma ray burst GRB 030329 looks like according to an artist. At 2.6 billion light-years from Earth, this is the closest GRB to us. Twin jets streak out from what was once a massive star that collapsed under its own weight into a black hole. The black hole's gravity pulls leftover material from the star toward itself. Then the black hole throws the material out at its poles, forming the jets. The jets are moving at nearly the speed of light and emit gamma rays. Slower material emits radio waves and visible light.

GRB 030329 was discovered in March 2003 by NASA's HETE-2 satellite, and then astronomers turned the VLA on it. They were able to determine the total energy output of the burst—optical, X-ray, and gamma ray telescopes had missed 90 percent of it. By comparing these results with those for other GRBs, the astronomers showed that although some bursts emit more energy in the form of gamma rays than other bursts, they are all the same kind of explosion.

entire rest of the universe for one or two seconds. Chances are against a GRB going off in our solar system's neighborhood, though.

The mystery of what was causing such energetic bursts remained unsolved for 30 years. And it was a big mystery: since the bursts were coming from all over the sky, they weren't just from something in our Milky Way galaxy. But if they weren't in our galaxy and were very far away, so much energy would be needed for them to be detected on Earth, they would have more energy than is possible as predicted by Einstein's equation $e = mc^2$ (energy equals mass times the speed of light squared). To say that Albert Einstein is wrong is a serious thing in astronomy. First, he hardly ever was. Second, the energy-mass-speed-of-light equation is one of the foundations of astronomy. If *that* equation turned out to be wrong, the entire universe might not be at all like astronomers thought.

Then on May 8, 1997, an Italian-Dutch satellite called BeppoSAX detected X-rays coming from the gamma ray burst GRB 9705089. Five days later, on May 13, astronomer Dale Frail turned the VLA to the burster and detected radio waves coming from it. From the VLA observations, the size of the GRB fireball could be determined and also the speed of its expansion after the explosion. One of the ways the VLA determined the size and speed of the fireball was by detecting its "twinkle." The way the VLA detects GRB twinkle is this. From Earth, stars appear to twinkle but planets don't, since they're big enough to be seen as disks, not points. Radio waves also twinkle (scintillate, actually)—but also only from point objects. If the twinkling stops, the object must have grown to disk size. For the May 1997 GRB to reach that size, it must have been bursting outward at close to the speed of light. It was just a tenth of a light-year across at first.

Because the VLA can "look" through dust, it has shown that the GRB explosions occur in dusty parts of galaxies, where young stars are likely to have formed. The VLA radio observations are sensitive enough to pinpoint the position of a faint GRB that optical astronomers might not see. Once the VLA had caught GRBs in the making, optical astronomers knew where to look for them. Then optical telescopes could be used to determine the GRBs' distance from Earth. The optical telescopes showed that gamma ray bursters are very far away, billions of light-years. Because light waves take so long to reach us from GRBs, we are really seeing far back in time. Now astronomers just had to come up with an explanation for how an object so distant could explode so energetically.

Armed with the VLA information, astronomers have come up with a possible cause for the bursts that would supply enough energy. Just one very massive star, much more massive than the sun, exploding at the end of its life, could release that much energy as it collapsed into a black hole. A big star collapsing like this is called a *collapsar*.

A GRB might develop like this. A massive star would form in a star nursery, a cloud of gas and dust. In just a million or so years the star would have used up all the fuel in its core. The star would then squeeze down into a black hole, sucking in any remainder of the star. Jets of material would shoot out of the hole, slamming into particles in space and generating the gamma ray blast we detect on Earth.

So a black hole is born explosively during a GRB—if anything so dark as a black hole can be said to be born—and a big star dies.

The radio galaxy 3C31 has gigantic, distorted plumes stretching 92,000 light-years out from its center.

FIVE

A Zoo of Galaxies

Stars don't randomly speckle the night sky. They gather in groups called *galaxies*. To study galaxies, which are huge groups of millions or billions of stars, astronomers must look far into space. But as distance from Earth increases, often dust in space does too, hiding what lies behind. The VLA telescope can "see" through dust, and so it has made many important discoveries about the most distant objects in the universe. The VLA has also served as a time machine, helping astronomers look into the long-ago past, when the super-energetic quasars formed.

Types of Galaxies

The millions or billions of stars in galaxies collect to form different shapes, and the force of gravity holds stars together in a galaxy. The sun and the stars you see at night are in the Milky Way galaxy. Some of those "stars" are actually faraway galaxies, filled with their own stars.

The spiral galaxy M33. The image is a composite made using the VLA, the Kitt Peak National Observatory telescope, and the Westerbork Synthesis Radio Telescope in the Netherlands. M33 is over 30,000 light-years across and more than 2 million light-years from Earth. It is part of the Local Group of galaxies, along with our own galaxy, the Milky Way.

The Milky Way galaxy is a grand-design spiral in shape and is about 100,000 light-years across. A *spiral* galaxy has curved arms flung out in space and a bulge at its center. The Milky Way galaxy has about six spiral arms. The sun is in the Orion arm, about 26,000 light-years from the galaxy's center. The sun and Earth aren't at the center of the galaxy or anywhere special, as far as we can tell.

Other galaxies are *elliptical*, either round or shaped like footballs. Still others are *irregular* in shape, like the Large and Small Magellanic Clouds, the galaxies closest to the Milky Way.

Then there are the *weird* galaxies. Weird galaxies are oddly shaped. Some weird galaxies have turned out to be two galaxies colliding.

Galaxies can also be *active*: something violent going on at their center makes them emit powerful electromagnetic waves. *Radio galaxies* are one kind of active galaxy.

Galaxies are often found in groups, and so is the Milky Way. About 30 galaxies belong to the Milky Way's group, called the Local Group. The group stretches about 30 million light-years end to end. Near one end is the Milky Way, and near the other is M31, also called the Andromeda galaxy. Andromeda is another large spiral galaxy. The Large and Small Magellanic Clouds are also part of the Local Group. The Large Cloud is 170,000 light-years from our galaxy, and the Small Cloud is 200,000 light-years away. A bridge of stars connects the two clouds, which orbit the Milky Way.

The Mysterious Center of the Milky Way

The center of our galaxy is shrouded in dust, and so visible-light telescopes can't see it. The VLA, on the other hand, can detect radio waves through this dust. In 1983 astronomers pointed the VLA at the exact center of the Milky Way. They found a spiral of hot gas coming out of it and then a distinct point of radio emission, Sgr A*, coming from the dead center of the galaxy.

Using the VLA, the astronomers watched this point for over 16 years to try to tell what they were dealing with. If they were detecting a small object, about the same mass as the sun, it would move quickly around the center of the galaxy, the way other stars close to Sgr A* were. But if they were seeing a massive black hole, it would seem barely to move. After all those years of observations, the astronomers could tell that Sgr A* is hardly moving, so it's probably a supermassive black hole, a gravitational monster holding the mass of millions of suns crunched together.

Because Sgr A* moves so little, part of what seems to be its motion is actually the motion of our solar system speeding around the center of the galaxy. From their measurements of Sgr A*, the astronomers could tell that the solar system orbits the center of the Milky Way in about 226 million years.

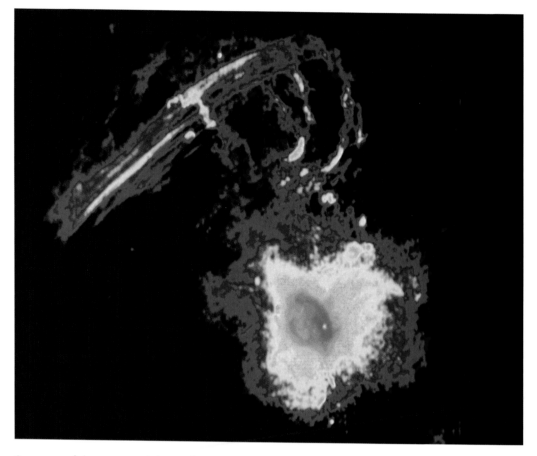

Structure of the center of the Milky Way galaxy. The center object is flinging out long, narrow streamers that the VLA detected at a wavelength of 20 centimeters. The streamers are pulled into these shapes by magnetic fields.

The strange events at the center of our galaxy have recently gotten even stranger. Astronomers did a computer analysis of the VLA data on the Milky Way's black hole, Sgr A*. They found that in the accretion disk around Sgr A*, where material is swirling down into the black hole, bubbles are forming every 106 days. The black hole seems to be boiling like a pot of soup. When the bubbles appear, Sgr A* emits more radio waves. The astronomers looked for objects around Sgr A*, like a star, that might be getting sucked in by the hole and creating the bubbles, but they found nothing. Astronomers are excited about the discovery of black hole bubbles because if they can figure out why the black hole is bubbling and why it bubbles every 106 days, they will have a better understanding of how black holes work.

Radio Galaxies

Often weird galaxies, which have unusual shapes, are active as well. Active galaxies are violent places, emitting tons of energy. (Our own Milky Way galaxy is just a regular spiral and is not an active galaxy.)

Radio galaxies, which are active galaxies, emit energy strongly as radio waves. The energy comes from the restless, churning center of the galaxy, where a supergiant black hole is active. Some radio galaxies emit powerful jets that beam out on either side of them. Jets form when gravitational energy from the black hole speeds up particles in the accretion disk around it.

Fornax A is a weird radio galaxy with shells and ripples. In this VLA image, white is visible light and shows an elliptical galaxy at the middle. The orange is radio waves and shows the giant lobes on either side of the galaxy. (Fornax A also has a small companion galaxy at the top right of the white part of the image.) Fornax A is thought to be what is left of two galaxies that collided 3 billion years ago. The collision may have resulted in a huge black hole at the center of the galaxy. (The black hole is shown by the lighter-colored circle in the middle of the white.) All of the particles and energy that created the two huge radio wave-emitting lobes came from the black hole.

The Antennae galaxy, NGC 4038/9, is two spiral galaxies that have collided. Green and white are optical light, and blue is the radio wave signal, from neutral hydrogen gas.

Then magnetic fields channel the particles into jets. The particles move away from the black hole at almost the speed of light. As the particles whip into space, they crash into particles already out there and pile up to form giant dumbbell-shaped lobes.

The radio galaxy Cygnus A is 600 million light-years away and one of the strongest radio sources in the sky. Cygnus A is thought to be a giant elliptical galaxy that recently swallowed a spiral galaxy. Evidence of this is a dust lane from the spiral that cuts across the middle of Cygnus A. When one galaxy "eats" another galaxy, it's called *galactic cannibalism*.

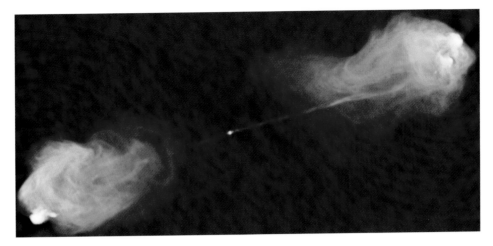

The radio galaxy Cygnus A in shades of orange, sporting two immense lobes connected by jets to the central galaxy (and black hole). The black hole is the small, bright circle at the center of the image. The image was made with the VLA in all four of its antenna configurations.

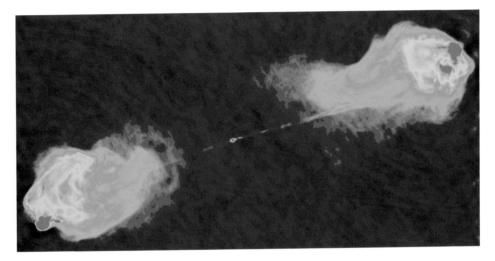

The radio galaxy Cygnus A in shades of blue, with red the brightest radio emission. Because radio telescopes don't actually see colors, the way the Hubble telescope and other optical telescopes do, astronomers can just assign colors to the radio signal they receive from a galaxy or other objects in space. Dark blue could stand for a stronger signal and light blue for a weaker one. The choice of color doesn't matter as long as other astronomers understand what is meant. If it's not totally clear, the astronomer will say under the image or in his or her research article what color scheme was used.

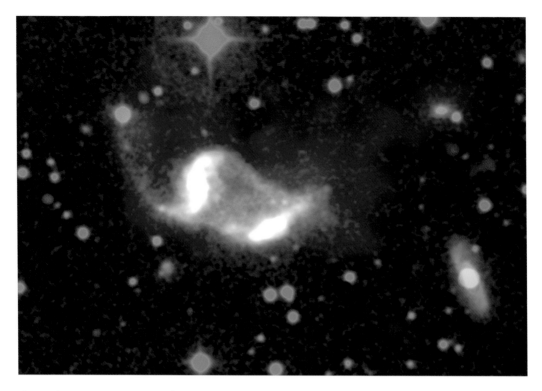

The galaxies UGC 813 and UGC 816 (white shapes) were normal galaxies that collided at a speed of about 500 kilometers per second 50 million years ago. Now the galaxies are moving apart, but they are still connected by magnetic fields and material torn from the galaxies. Visible light is green, neutral hydrogen gas is blue, and radio emissions are red.

You won't be surprised to learn that the VLA, a radio telescope, has been used to study radio galaxies—big time. One long-term study focused on the powerful radio jets that beam out of radio galaxies.

Like so many other astronomical projects, the study of the radio jets involved different astronomers with different kinds of telescopes. More than 20 years ago astronomer Frazer Owen, using the VLA, began a radio survey of 500 galaxy clusters. The VLA showed that one of the galaxies had a jet. This galaxy, named 0310-192, is almost a billion light-years away. So far, so good. But then an image with the optical telescope at Kitt Peak National Observatory showed that 0310-192 might be a spiral galaxy. A detailed image made with the Hubble telescope proved it.

Spiral galaxies were thought to be the wrong kind of galaxy to form a jet. Usually elliptical galaxies or galaxies that are colliding form jets—the pressure from the material between the elliptical galaxies or colliding galaxies keeps the jets from falling apart. But here was a spiral with a jet,

A combined VLA and Hubble telescope image of the galaxy 0313-192. The galaxy is seen sideways, or edge on. If you were looking down on the galaxy, it would look just like the spiral galaxy seen above and to the right in the image.

in a looser collection of galaxies called Abel 428. So astronomers had to rethink what was happening with the jets.

A closer look at 0313-192 revealed that the galaxy is twisted. That may have been caused by a close encounter with another galaxy, or 0313-192 may have swallowed a companion dwarf galaxy. So 0313-192 isn't just a simple spiral. Its unusual structure could make possible the formation of the jet.

When Black Holes Smash

If galaxies have black holes at the center, what happens to each galaxy's hole when two galaxies collide? Astronomers haven't yet found two black holes at the center of any galaxy. The VLA has produced many images of radio galaxies showing the jets of radio wave-emitting particles that shoot out of spinning black holes at the galaxies' centers. In about 7 percent of radio galaxies, those jets have flipped—which could mean that the black holes in those galaxies suddenly moved. Black holes have swallowed huge amounts of matter, and to move one in space would take tremendous force. The only thing that could whack a black hole into a new position is another black hole.

As the two black holes of separate galaxies draw near to each other, they spin faster and faster and finally crash. These collisions could happen once a year in the universe.

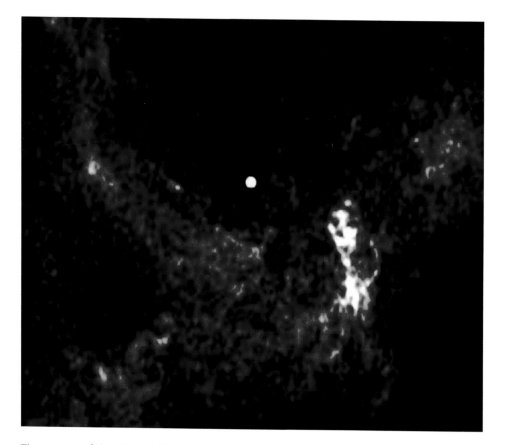

The center of the Milky Way galaxy, seen close up with the VLA. The bright white dot shows Sgr A*, the exact center of the galaxy. Sgr A* is a massive black hole.

The Monster Engines of Quasars

By the early 1960s, astronomers had found a group of odd radio sources in the sky. They looked like stars—sometimes like very faint stars. In 1962 a British astronomer, Cyril Hazard, tracked the source of the radio waves from one of these radio sources to a star called 3C 273, in the constellation Virgo. But this so-called star was blasting out radio signals, and even the regular, visual light didn't look like the light coming from other stars.

Finally in 1963 Maarten Schmidt at Mount Palomar Observatory figured out what these strange "stars" were. The light coming from them was normal, just coming from very far away—3C 273 was 3 billion light-years away and moving away from Earth at almost 48,000 kilometers per second (almost 30,000 miles per second). But 3C 273 was giving off way too much energy to be a star. Astronomers called the new objects *quasars*, short for *quasi-stellar radio sources*. Thousands of quasars have now been found.

Red- and Blue-Shifted Light: The Doppler Effect

Astronomers didn't recognize the light from 3C 273 at first because the light was *red shifted*. The frequency of light can be shifted in the same way the frequency of a sound can. For example, when you hear the horn of an approaching train, the sound of the horn becomes higher pitched as the train moves closer to you. After the train passes you, as it moves away, the sound of its horn is lower in pitch, seeming to wail mournfully. The frequency of light becomes lower if an object is moving away from you and becomes higher if the object is coming toward you. Light shifted to higher frequencies is said to be *blue shifted*. Light shifted to lower frequencies is red shifted.

The light from most quasars is extremely red shifted, which means that the quasars are moving away very fast from Earth. They are also very far away. Because they are so distant, when we look at them, we see far back in time—their light has taken billions of years to reach us. The universe may have been very different so long ago. That complicated astronomers' efforts to explain what quasars are. So they aren't stars. What, then?

The Hubble telescope provided some answers. Quasars live in galaxies—the galaxy surrounding the quasar is just much dimmer than the quasar and harder to see. Quasars seem to live in various kinds of galaxies, some normal, some colliding with neighboring galaxies. For such

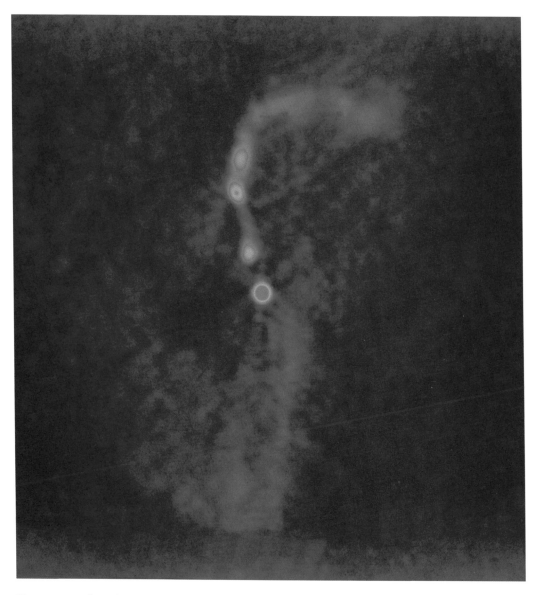

The quasi-stellar object (quasar) PKS B2300-189. The bright object at the center of the image is the actual quasar. The other bright objects and blue "tails" tracing an S shape on the sky are radio wave-emitting material ejected from the quasar.

powerful objects, quasars are quite small, only a couple of light-months across. (The Milky Way galaxy is 100,000 light-*years* across.)

How can quasars emit so much energy? Only a black hole could do it. A quasar turns on when gas and stars fall into a black hole. As the black hole feeds, it produces the enormous amounts of energy coming from the quasar. So a quasar is thought to be the bright, active center of a galaxy.

Seeing far into the past, astronomers using the VLA have studied the bright quasar galaxy APM 08279+5255. The quasar galaxy is now 12 billion years old, but since the light took 12 billion years to get to Earth, we're seeing the galaxy when it was young. APM 08279+5255 has the required hungry black hole at its center, and the galaxy teems with newly forming stars—the VLA detected enough gas to make 100 billion suns. This blast from the past will help astronomers understand how galaxies form and live out their lives.

Quasars don't seem to form much these days. Astronomers think they may have evolved into something else. Normal galaxies, like the Milky Way, have a huge black hole at their center, but those black holes may have mostly run out of fuel: stars and dust to eat. Like people, galaxies may settle down in their old age.

Sometimes, though, quasars return to their youth. Astronomers used the VLA to study three nearby quasars (*only* 630 to 830 million light-years away). Observations with visible light had shown that one of the quasar galaxies was colliding a bit with another galaxy, the second quasar galaxy had a neighbor galaxy but the two galaxies seemed to be living together peacefully, and the third quasar galaxy looked like it was alone in space. The VLA showed, though, that the gas of all three galaxies had been disturbed by a collision with another galaxy. More VLA studies of galaxy collisions should help show how these collisions turn on quasars.

The Big Bang of the Universe

In the beginning of the universe was . . . what?

A shattering, sizzling fireball, expanding everywhere in space 13.7 billion years ago. The fireball is thought to have been at a temperature of 10^{19} degrees Celsius (10 followed by 19 zeros) at 10^{-32} second (0.00000000000000000000000000000001 second) after the very start of time.

The fireball, containing everything known in the universe, quickly cooled, in about a hundred seconds after the beginning of time, to 10^9 degrees Celsius, and hydrogen and helium formed out of lighter particles. Hydrogen and helium are the main elements that make up the sun and other stars.

After about 200 million years, the universe had cooled to just 50,000 degrees Celsius (almost the same in degrees Fahrenheit) and the very first stars began to shine, lighting up the black of space. But no one was around to see it—Earth wouldn't form for billions of years, and people wouldn't evolve for billions of years after that. The galaxies continued to expand, moving away from each other like polka dots on the surface of an expanding balloon.

Is there any proof that this is what happened? In 1965, Arno Penzias and Robert Wilson, who were scientists at Bell Laboratories, wanted to do a study of radio signals coming from the Milky Way galaxy. They found an annoying radio noise that wasn't just coming from the Milky Way, but from everywhere in the sky. The electromagnetic waves they detected were at the right wavelength and temperature to be left over from the big bang. For their discovery, called the cosmic background radiation, Penzias and Wilson shared the 1978 Nobel Prize in physics.

Will the universe expand forever, or will galaxies and black holes and everything else be pulled back together by the force of gravity, eventually smashing in a "big crunch"? A NASA spacecraft, the Wilkinson Microwave Anisotropy Probe (WMAP), has come up with some answers to this question. WMAP, orbiting around a point called L2 (Lagrange Point 2), about a million miles from Earth, found evidence that the universe consists of 4 percent ordinary atoms, like hydrogen, 23 percent some kind of unknown *dark matter*, and 73 percent *dark energy*, which seems to act like antigravity. All of those put together seem to be enough to battle the

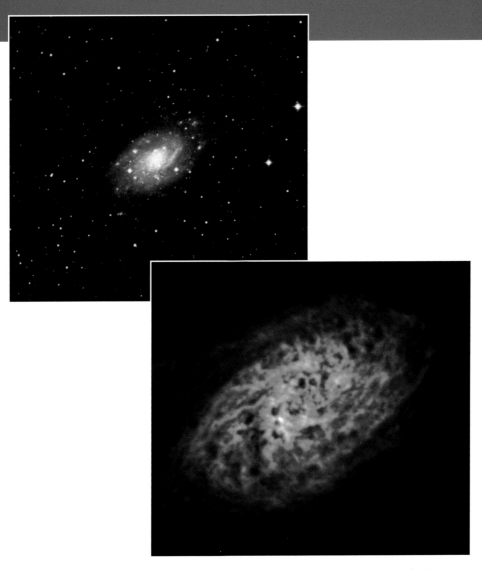

The spiral galaxy NGC 2403 appears quiet in the image at the top, which shows the galaxy in visible light. But the radio image of NGC 2403 at the bottom reveals a huge cloud of hydrogen gas pockmarked by violent supernova explosions. The galaxy takes up as much space across the sky as the full moon.

force of gravity and keep the universe expanding, flinging people, planets, and stars deeper into space. Astronomers have found that instead of slowing with age, the expansion of the universe is speeding up. But a new theory suggests that the universe has existed forever and the big bang is just a moment in that long existence.

Accompanied by the moon, the VLA searches the sky.

SIX

The Future of Radio Astronomy

Another Radio Telescope: The Very Long Baseline Array

The Very Large Array radio telescope has 27 antennas, so that the telescope is like one big antenna combined. Especially when the antennas are spread out as far as they will go, in the A configuration, the telescope has good resolution—it can see details well. But suppose the antennas were spread out even farther apart. Then the resolution should be even better.

The Very Long Baseline Array, or VLBA, cost $85 million to build and began to make observations in May 1993. Like the VLA, the VLBA is a radio telescope. It consists of 10 antennas that are 25 meters (82.5 feet) in diameter, the same diameter as the VLA antennas. But eight of the VLBA antennas are spread out across the entire continental United States. One more antenna is in Mauna Kea, Hawaii, and one is in St. Croix in the U.S. Virgin Islands, south of Florida. The VLBA can produce images that are as detailed as if it were a single antenna 8,600 kilometers across (5,160 miles across).

The 10 antennas of the VLBA stretch across the continental United States and then some. The antenna at the far right is in St. Croix in the U.S. Virgin Islands, and the one at the far left is in Mauna Kea, Hawaii.

Radio signals detected by the VLBA are taped by a tape recorder at each antenna and sent to the VLBA correlator in Socorro, New Mexico. A computer doing 750 billion operations a second correlates the tapes, correcting them for differences in the antennas' position and the time that the signal was received. Then astronomers can turn the results into images. The VLBA can also communicate with radio telescopes in other countries and satellites in orbit.

Astronomers have made some startling discoveries with the VLBA. Recently they found a black hole, not sitting still at the center of its galaxy but speeding crazily through the Milky Way galaxy, dragging a star along with it. This black hole, named XTE J1118+480, sucks in material from its companion star as the hole and star travel—eating on the run. The VLBA could detect the movement of the two because the telescope can see such fine details.

What makes the black hole even more interesting is that it's a *micro-quasar*. It's spewing out jets of particles, which then emit radio waves, just like some other quasars, but it's only 6,000 light-years away—most quasars are billions of light-years away—and it's small for a quasar, only the size of a massive star. Astronomers think the black hole formed when a very ancient star exploded in a supernova explosion, even before the Milky Way galaxy existed. Then a kick from gravity pushed the hole out of the star cluster where it formed. The black hole grabbed a nearby star, and for the past 230 million years both hole and star have traveled together. Hundreds of thousands of these old black holes are thought to be traveling around the Milky Way.

The same astronomers who discovered the out-of-control black hole found a neutron star doing the same thing with a companion star of its own. The two stars are called Scorpius X-1, and they have been traveling together for about 30 million years, shooting in and out of the Milky Way. Right now they are about 9,000 light-years from Earth.

Here's another black hole story, but about a bigger hole—the VLBA has also been turned on Sgr A*, the black hole equal to almost 4 million times the mass of the sun that lurks at the center of the Milky Way galaxy. The astronomers expected Sgr A* to be symmetrical, which means if you cut it in two, you'd get equal-shaped halves. But the VLBA revealed that just outside the actual black hole, Sgr A* has a radio wave-emitting area that is shaped like a cigar. (Nothing can be seen or detected in the actual black hole—remember, the hole's gravity pulls in light and radio waves.) None of the current theories for how black holes work can explain the cigar. A jet or wind coming out of the hole might help explain it. So once again, astronomers will adjust their theories to fit the new observations.

With the VLBA, astronomers have improved their measurements of the size of Sgr A*. Astronomers now think that the actual black hole is about 14 million miles across, which means it would fit inside Mercury's orbit. That's a very small space for an object with a mass of almost 4 million suns and means that the total mass of the black hole is extremely squashed down, or concentrated.

The VLBA has also built on the VLA's work on planetary nebulas. The VLA found one old star, K3-35, that had just begun its planetary nebula stage, when the star throws a shell of gas into space as it dies. The VLBA detected another such star, W43A, and in doing so helped solve a mystery about planetary nebulas. Many planetary nebulas have strange shapes. But if you think about it, it's odd that a sun-shaped object, which is basically round, can blow up into a planetary nebula that isn't round at all.

The star W43A is about 8,500 light-years from Earth. This star is at the end of its lifetime, and astronomers think it will soon form a planetary nebula. Using the VLBA, astronomers found that the old star is spewing jets of water molecules like a garden hose full of water. Whatever is squirting out the jets seems to be slowly rotating, twisting the jets. These spinning jets of water could produce the shapes of planetary nebulas.

The water jets are probably only a few decades old. When the star collapses into a white dwarf, ultraviolet radiation will rip the jets apart, so astronomers were lucky to catch the star during this short part of its lifetime.

A New and Improved VLA: The Expanded Very Large Array

The VLA telescope has been updated, but some of the electronics equipment dates to the 1970s—PCs weren't even around then. This equipment takes up a lot of room, because computers and other electronics had to be bigger in the old days. Scientists are upgrading the VLA in two stages. In the first stage, improvements include installation of new electronics and a new correlator, and the old waveguide cables that carry signals in from the antennas are being replaced by fiber optic cables—100 times more information can travel down a fiber optic cable. In the second stage, eight new antennas will come online, linked to the old 27 antennas and to the closest VLBA antennas. The new antennas will still be in New Mexico but as far as 250 kilometers (150 miles) from the old group. The new, improved VLA will be called the Expanded Very Large Array, or EVLA. It will be able to detect many more radio wavelengths than the old VLA.

The EVLA will be able to do everything the VLA did, only better. For example, right now, the VLA can see only about a third of gamma ray bursters. The EVLA should be able to see them all.

So what will the next major discovery in radio astronomy be? It's hard to say exactly. Ten people have won Nobel Prizes in physics for work on astronomy, and of those 10, six won the prize for work with radio telescopes. Future discoveries in radio astronomy are sure to be bizarre, extraordinary, and out of this world.

Glossary

accretion disk: The ring of material swirling around a black hole, pulled toward it by the black hole's gravity.

active galaxy: A galaxy that emits an unusual amount of energy, generally radio waves, X-rays, gamma rays, or all of these.

antimatter: The opposite of matter. An antimatter universe would consist of antiprotons, antineutrons, and antielectrons. If these particles met up with their counterparts in our universe—protons, neutrons, and electrons—they would annihilate each other. See also *matter.*

atom: The smallest unit of an element, like hydrogen or helium, that can undergo the reactions typical of that element, like hydrogen reacting with oxygen to form water or hydrogen fusing with itself to form helium in the sun's nuclear reactions. An atom is made up of protons and neutrons bound into a nucleus at the center, with electrons orbiting the nucleus. See also *electron, neutron,* and *proton.*

big bang: The explosion of the universe into being at the beginning of time. The hot fireball cooled off and expanded, eventually forming stars, galaxies, and planets.

black dwarf: The cinder of a burnt-out star. The sun will turn into a black dwarf when it has exhausted its nuclear fuel and completely cooled off. See also *white dwarf.*

black hole: What is left of a supermassive star when it reaches the end of its life and blows off its outer layers. The core collapses until it is so tightly squeezed, its powerful gravity keeps matter and even light from escaping. Black holes can form at the heart of galaxies.

blue shift: A change in electromagnetic waves to higher frequencies or shorter wavelengths.

collapsar: A massive star that collapses into a black hole, emitting gamma rays, radio waves, and visible light.

dark energy: What seems to be empty space may hold a strange new type of energy that acts as antigravity, pushing the universe apart.

dark matter: An astronomical object or particle, like a planet or a black hole, that can only be detected when its gravity affects other objects.

diameter: The length of a straight line through the center of a circle. The diameter of Earth is the distance straight through it, from one side to the other.

eclipse: A solar eclipse occurs when the moon passes in front of the sun, blocking the sun's light. A lunar eclipse occurs when Earth passes between the sun and moon and hides the moon.

electromagnetic spectrum: The entire range of wavelengths or frequencies of electromagnetic waves.

electromagnetic wave: Radiation, including radio waves, visible light rays, X-rays, and gamma rays, that travels at the speed of light. The different electromagnetic waves have different wavelengths or frequencies depending on their energies.

electron: A light particle that, along with protons and neutrons, makes up an atom, the smallest unit of an element like hydrogen. An electron has a negative, or minus, charge. Electrons orbit the neutrons and protons of an atom, which are bound together.

event horizon: The limit to where light or matter can escape from a black hole. The size of the black hole determines how far out its event horizon will be.

frequency: The number of electromagnetic waves that pass a point per unit of time.

galactic cannibalism: When galaxies collide and one swallows the other.

galaxy: A big collection of stars, gas, and dust held together by gravity. Galaxies can be spiral, with a central bulge and arms (like the Milky Way), elliptical (either round or like a stretched-out circle in shape), or irregular (like the Large and Small Magellanic Clouds). Weird galaxies have distorted shapes and may result from galaxies colliding.

Large and Small Magellanic Clouds: The galaxies closest to the Milky Way. They are irregular in shape. The Small Cloud is 200,000 light-years from the Milky Way and the Large Cloud is 170,000 light-years away.

light-year: The distance light travels in one year, 9.6 trillion kilometers (9,600,000,000,000 kilometers) (6,000,000,000,000 miles). Light and all other electromagnetic waves move at 300,000 kilometers per second (186,000 miles per second).

Local Group: The Milky Way is one of about 30 galaxies that are fairly close to one another. They are all part of the Local Group.

mass: The property of an object that causes it to have weight in a gravitational field and to resist change in its motion.

matter: Particles or groups of particles. Our universe is made up almost completely of matter—protons, neutrons, and electrons, which put together form atoms—but a mirror-image universe of antimatter could exist. See also *antimatter.*

NASA: National Aeronautics and Space Administration. This U.S. government agency is in charge of space flight, including the space shuttles, satellites, and unmanned spacecraft.

nebula: A cloud of gas and dust. The Orion nebula, which is 1,500 light-years from Earth, is a birthplace of stars.

neutrino: A tiny particle with no electric charge and either no mass or very little mass. Neutrinos are given off in the nuclear reactions in the sun.

neutron: Neutrons, along with protons, make up the nucleus of an atom. (The other part of an atom is the electrons that orbit the nucleus.) Neutrons have no electric charge. See also *atom, electron,* and *proton.*

neutron star: A star made up mostly of neutrons, squashed together tightly by gravity. Neutron stars are very small, only about 10 to 20 kilometers (six to 12 miles) in diameter. These stars are the core that is left at the end of some big stars' lives, after the explosions that blow off the stars' outer layers. See also *pulsar.*

photon: Electromagnetic radiation, such as visible light, can be looked at as electromagnetic waves, with a certain wavelength and frequency, or as photons, particles of energy.

planetary nebula: The hot shell of gas blown off a star like the sun near the end of its life as the star collapses into a white dwarf. Radiation from the dwarf makes the shell glow. See also *white dwarf*.

positron: The opposite of an electron. Positrons are lightweight particles given off in the sun's nuclear reactions. In an antimatter universe, positrons would take the place of electrons. See also *electron*.

proton: Protons, along with neutrons, make up the nucleus of an atom. (The other part of an atom is the electrons that orbit the nucleus.) Protons have a positive electric charge. See also *atom*, *electron*, and *neutron*.

pulsar: A rapidly spinning neutron star that strongly emits electromagnetic waves, including radio waves, light waves, and X-rays. See also *neutron star*.

quasar: A bright, very energetic source of light and radio waves thought to be the active center of a galaxy. Most quasars are billions of light-years away, so we see them as they were long ago in the past.

radio galaxy: A galaxy that emits large amounts of radio waves.

red giant: The bloated, cooler stage in the life of a star like the sun, when it has used up most of its nuclear fuel.

red shift: A change in electromagnetic waves to lower frequencies or longer wavelengths.

solar system: The sun, nine planets, moons, comets, and asteroids. Other solar systems may exist around other stars.

supernova: The explosion of a star much bigger than the sun that has reached the end of its life. Supernovas are extremely bright— suddenly the star may appear hundreds of millions of times brighter than it was before the explosion.

superwind: The blast from the surface of an unstable, dying star.

wavelength: The distance from peak to peak in waves.

white dwarf: The hot remnant that remains of stars that were once about the size of the sun but exhausted their nuclear fuel and blew off their outer layers. The white dwarf stays hot for a while with heat left over from its nuclear reactions. Eventually, the star will cool off completely and turn into a cold cinder, called a *black dwarf*.

Illustration Credits

Index

Italic numbers indicate illustrations.